CW00602874

WORKMANSHIP STANDARDS MANUAL

Quality Assurance

Company name/stamp

Approved by ——————————————————————

Date ——————————————————————————

Every effort has been made to ensure the typographical and technical accuracy of the contents of the manual. Should any errors or discrepancies be found the publishers will be pleased to receive any constructive comments for consideration in future editions.

WORKMANSHIP STANDARDS MANUAL

Quality Assurance

Ray Skipp (MIQA)

Third Edition
revised by
Andrew Heydn

Blackwell
Science

First edition published by Capritone 1986
Second Edition published by BSP Professional Books 1989
Reprinted 1991, 1993
Third Edition published 1997

Set in 11/12.5 pt Ehrhardt
by DP Photosetting, Aylesbury, Bucks
Printed and bound in Great Britain by
Hartnolls Ltd, Bodmin, Cornwall

DISTRIBUTORS

Marston Book Services Ltd
PO Box 269
Abingdon
Oxon OX14 4YN
(*Orders:* Tel: 01235 465500
         Fax: 01235 465555)

USA
Blackwell Science, Inc.
Commerce Place
350 Main Street
Malden, MA 02148 5018
(*Orders:* Tel: 800 759 6102
           617 388 8250
        Fax: 617 388 8255)

Canada
Copp Clark Professional
200 Adelaide Street West, 3rd Floor
Toronto, Ontario M5H 1W7
(*Orders:* Tel: 416 597-1616
           800 815 9417
        Fax: 416 597 1617)

Australia
Blackwell Science Pty Ltd
54 University Street
Carlton, Victoria 3053
(*Orders:* Tel: 03 9347 0300
        Fax: 03 9347 5001)

A catalogue record for this title is available from the
British Library

ISBN 0-632-04003-3

Library of Congress
Cataloging-in-Publication Data
is available

Objective

To produce high quality products.

Scope

To provide a basis of manufacturing standards throughout manufacturing organisations.

Should these standards conflict with those of a specific customer's contract then the customer's standards will prevail.

Notes to Reader

1. This document is divided into Sections identifying principal subjects.
2. Each Section is further divided into Subsections embracing specific subjects within the main heading.
3. Weights, measures and dimensions are given in metric units.

Credits

Contributions on surface mount are by Mr Bob Willis of Dimension 2 Technology, Reading.

Figures 2.35–2.60 are reproduced courtesy of Dimension 2 Technology.

Figures 7.13–7.29 are reproduced courtesy of GEC/EITB.

Contents

SECTION 1: WIRING

A – Colour codes

NOTE: The following colour codes identify conductors used in direct current (DC) and alternating current (AC) systems.

1. INTERNAL UNIT WIRING

DC Circuits

- Positive – red
- Negative – black
- Earth – green/yellow

One-phase AC circuits

- Line – red
- Neutral – black
- Earth – green/yellow

Three-phase AC circuits

- Phase 1 – red
- Phase 2 – yellow
- Phase 3 – blue
- Neutral – black
- Earth – green/yellow

For other types of circuits, see the latest IEE Wiring Regulations.

2. EXTERNAL FLEXIBLE CORDS AND CONNECTING CABLES

One-phase AC circuits

- Line – brown
- Neutral – blue
- Earth – green/yellow

Three-phase AC circuits

- Phase 1 – brown or black
- Phase 2 – brown or black
- Phase 3 – brown or black
- Neutral – blue
- Earth – green/yellow

The three brown (or black) phase leads will be identified, in the order in which each attains maximum potential, by the application at each end of the insulation of the conductor of a white sleeve or binding of proofed tape bearing black indelible Arabic numerals 1, 2 and 3, respectively.

DC Wire Circuits

- Positive – red
- Negative – black
- Earth – green/yellow

B – Types of Wire

1. RESPONSIBILITY

It is the responsibility of the design authority to specify exactly the type of wire to be used for all applications.

2. ACTION

The wire type to be used must be shown clearly on the applicable drawings and parts list. There must be no deviation from the designated type without the prior consent of the design authority and the knowledge of the quality manager.

C – Identification of wires

1. IDENTITY

All interconnecting wires must be easily identifiable at both terminations of individual wires to facilitate inspection and service.

2. METHOD

All interconnecting wires run either singly or in

groups (i.e. cableforms, etc.) and must be identified in one of the following ways:

- Single wires easily visually traced throughout their entire run from one termination to the other
- Each wire of a complex group or cableform identifiable at both of its ends by the fitting of a numbered and/or colour coded sleeve or sleeves. The marker sleeves to be fitted so that the code is read left to right from the outer end of the wire
- Each wire of a complex group or cableform having a different colour or colour combination of insulation to every other wire in its group

Material used for cable markers shall be compatible with the environment that the equipment is intended to be used in.

3. COAXIAL, SCREENED & MULTICORE CABLES

Single coaxial or screened cables, multicore cables or groups of wires bound together as a free cable or cableform and terminating at one or both ends in a plug, socket, fanning strip or terminal block must have such termination separately identified.

D – Routing of Wires

1. APPEARANCE

Unless it is a design requirement to be otherwise, all wires must be routed in such a way that the finished product will present a neat, professional appearance.

2. ROUTING

Where the design permits, wires must be routed parallel with the chassis sides of printed circuit board (PCB) edges.

3. CRITICAL PATHS

Where a design requires wires to be routed by a critical path, clear and explicit instructions must appear on the relevant drawings and there must be no deviation from this requirement.

4. PLANNING

Proposed wire and cableform routes must be carefully planned to avoid wires being trapped under screws, components, access covers, etc.

5. PRINTED INFORMATION

Wiring must not mask printed information.

6. HEAT PRODUCING COMPONENTS

Wiring must be routed well clear of all heat producing components and must not restrict areas provided for ventilation or air ducting.

7. SITING

Wires must not pass or obscure:

- Fixing screws or removable parts or modules
- Apertures through which adjustments are made
- Sharp metal edges or protrusions (without adequate protection against chafing of the wire insulation)
- Moving parts such as switches, spindles, shaft couplings, fans, etc.

8. LENGTH

Wires must be sufficiently long to lay in to the designated route without cutting corners or stretching around component bodies, etc.

9. SERVICING COMPONENTS

Wires to items which may have to be 'lifted' for servicing (e.g. PCBs) should be routed so that the item may be hinged upwards from one side without undue strain on the wires.

10. SUPPORT

Long lengths of unsupported wire or cableform must be avoided except for wires specifically provided to enable a unit, module or part to be withdrawn without disconnection of the wires.

11. STABILITY

Wires grouped into cableforms must be secured so that they cannot move their position in the equipment due to vibration or movement of the equipment.

12. SUPPORT SPACING

All cableforms must be fixed and supported at intervals along their length not exceeding 12 times the diameter of the cableform. The point where the cableform terminates, termination of short branches, and passage of the cableform through a grommeted hole, may all be deemed to be 'fixing and support points' but if no such supports exist inside the terms of the 12 × diameter rule then the cableform must be secured, by the use of clamps or Tyraps to a standing part of the chassis or frame.

Unprotected metal clamps must not be used for securing wires or cableforms.

All wire or cableform fixings must be tight enough to hold the wire or cableform securely against random movement but not tight enough to distort or damage the insulation.

13. AREAS WITH HINGES

When a cableform has to pass into a hinged section of an equipment the cableform must be routed so that the axis of the cableform is kept as near parallel to the axis of the hinge as possible. Tortional rotation, i.e. twisting, of the cableform is preferable to right angle bending with operation of the hinged section. The cable should not chafe or foul other cables or components when the hinged section is operated.

14. HEAT SOURCES

(Polyvinyl chloride (PVC) insulated wires and multicore cables must always be routed to be as far away as possible from sources of heat. The softening region for PVC is $75°-85°C$ and continuous exposure to elevated temperature causes deterioration of the plastic material with consequent loss of properties. PVC insulated cables (especially coaxial cables) must never be fitted in a location where they will be subjected to temperatures continuously higher than $55°C$.

15. APPROACH TO TERMINATION POINTS

Single wire

Except for cases where it is necessary for technical reasons to run wiring by the shortest possible route or when short lengths of preformed wire are laid in between two terminating points, wiring should always approach termination points in an easy semi-circular sweep to allow for easy disconnection and reconnection during service and to allow enough wire to reconnect should the soldered end of the wire be accidentally broken off.

Cableforms

In the case of cableform terminating wires it is essential that the wire should not be laid straight from the breakout point in the cableform to its soldered termination because a breakage at the wire end would then require the whole cableform to be disturbed to replace the broken wire. On the other hand, excessively large loops of wire are unnecessary and look untidy.

16. CABLEFORM TERMINATIONS

It is not intended to specify rigid standards of length for cableform terminating wires and the following dimensions must be treated purely as a guide. If the quoted dimensions are followed approximately and the terminating wire is formed into a smoothly curved loop between the cableform breakout point and the terminating point (after the joint has been made) then a high degree of uniformity will be achieved and sufficient wire will still be available to provide a further 10 mm (approx.) stripping for any subsequent repair which may be required.

17. WIRE LENGTH AT CABLEFORM TERMINATION

The following instructions apply to cableform wires which are not more than 3 mm overall diameter.

To calculate the length of wire required to form a standard loop between the breakout point of a wire leaving a cableform and its soldered termination point:

(1) Measure the direct line between the breakout point and the termination point. This will be called dimension X.

(2) If dimension X is between 13 mm and 77 mm add 19 mm to this length. The total length of X + 19 mm will be the length of the insulated lead required to form an acceptable sized, approximately semicircular loop between the cableform and the terminating point. A further length of at least 14 mm should be allowed for bare conductor used to make the soldered joint.

E – Cableforms

1. METHOD OF CONSTRUCTION

Where two or more interconnecting wires in equipment follow the same route for a large part of their run they shall be fastened together into a cableform, provided that this will cause no technically detrimental effect. This is desirable for several reasons:

(1) Consistency and ease of wiring
(2) More secure and reliable than individual wiring
(3) Reduces wiring costs
(4) Improves tidiness and general appearance

Cableforms must be made to a plan on a pegboard with each wire laid in a specified sequence.

It is permissible for an experienced wire person to lace up networks of separate wires into a cableform as the wires are fitted. This method of cableforming is generally not very satisfactory (it only truly meets item (4) above) and should only be used where a cableform could not be fitted into a particular piece of equipment, such as when preformed cables are required to be formed in three dimensions.

All wires in a cableform should lie parallel with one another and not cross other wires in the form except where it is necessary to bring an internal wire out of the form.

Soldered joints are not permitted in the laced portions of cableforms in newly manufactured equipment. Should a wire be damaged during the manufacture or fitting of a cableform then the whole length of wire must be replaced.

During the servicing or repair of equipment, to reduce the cost of the repair a damaged single ordinary wire (solid or stranded conductor) in a cableform may be repaired in the following manner:

■ Cut out the damaged portion of wire at least 38 mm on either side of the damage
■ Fit a 19 mm long heat shrink sleeve of appropriate size on each of the cut ends of the wire left in the cableforms.
■ Strip the insulation of the cut ends 10 mm
■ Using wire similar in all respects to that being repaired, cut, strip and form a piece of wire just long enough to connect the cut ends of the damaged wire
■ Solder the new cable into place using lay on joints, taking care not to damage the other cables
■ Push the sleeve over the bare joints after inspection of the joints and heat sleeve until it shrinks to a snug fit over the cable
■ When complete, inspect the sleeve for any break through

NOTE: This type of repair should be limited to one per cableform and approval of the quality manager must always be obtained before proceeding with the repair.

Mains wiring must be formed into a separate cableform from all other wires and must always conform to the current IEE Regulations.

There must be no damage to the wires or to the insulation in the cableform in newly manufactured equipment.

There must be no strain on individual wires at 'T' junctions or 'take out' points.

There must be no kinks in the wire at any point. Wires in a cableform may be secured by any of the following means:

■ Lacing with a continuous tie using PVC covered nylon cord or waxed nylon braid
■ Individual strapping using nylon 'Tyrap' or perforated PVC strip and studs
■ Black PVC continuous sleeving
■ Expandable braided plastic cable sleeve
■ Hellerman spiral binding
■ Slit harness wrap
■ Twist locks

The cableform drawing shows a plan view of the cableform and is used as a template to set up the loom. The drawing indicates the positions in which the cableform pins should be placed and the positions of the ends of the wires. The template is divided into wire outlet zones for convenience of reference.

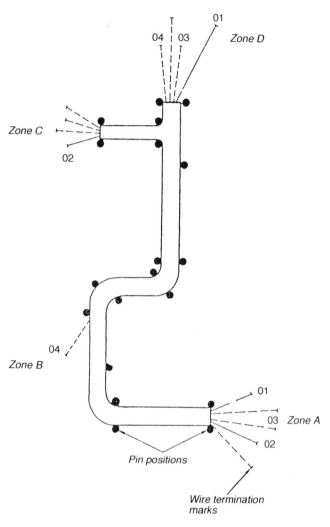

Fig. 1.1 Cableform template drawing.

The run out sheet indicates the order in which the wires are laid. It also gives the zone in which each wire end is located. For instance, the pink wire 01 (with black and brown markers) goes from zone D to zone A on the cableform drawing.

Run out table					
Wire		Zone			Length
no	Colour	To	From	Colour	cm
01	Brown/black	A	D	Pink	40.0
02	Brown/blue	A	C	Pink	39.0
03	Brown/red	A	D	Pink	40.0
04	Brown/green	B	D	Pink	26.5

Fig. 1.2 Run out sheet.

Lay out the cableform drawing on the board. Taking care not to stretch the drawing by excessive smoothing.

Fasten the edges of the drawing down with draughting tape.

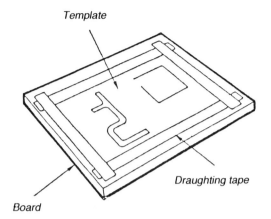

Fig. 1.3 Preparing the pinboard.

Staple the springs in position whenever wires come out from the main cableform as follows:

- Place the spring in position and staple one end
- Use the hammer to drive the staple into the board
- Stretch the spring so that a wire can fit securely between each coil and staple the other end of the spring

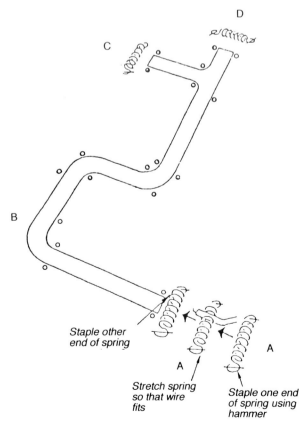

Fig. 1.4 Placing the springs.

Using the hammer, knock in the cableform pins in the positions shown on the drawing. When a large bend radius is required, use several pins to ensure that the radius is correct.

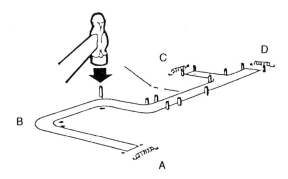

Fig. 1.5 Fixing the pins.

Preparing the wires – example

(1) Cut a 40 cm length of wire
(2) Strip each end 14 mm and twist slightly to avoid whiskering
(3) Place black and brown markers at each end of the wire. Take care to position markers the correct way round
(4) The wires can either be cut, trimmed, markered and laid in the cableform one at a time, or all wires can be cut, trimmed, and markered before cable forming, and then selected and laid in the position required.

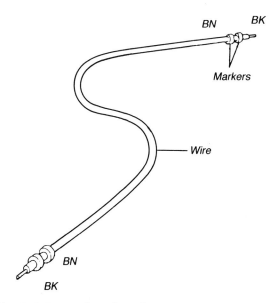

Fig. 1.6 Preparing the wires.

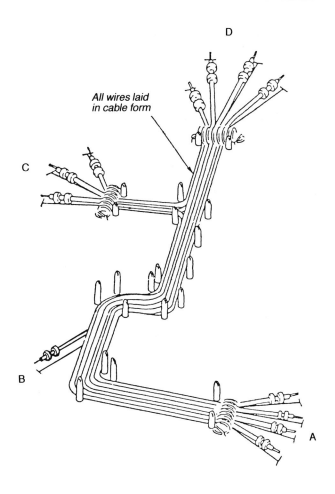

All wires laid in cable form

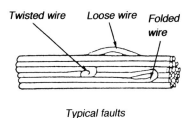

Twisted wire   Loose wire   Folded wire

Typical faults

Fig. 1.7 Wires in position before strapping.

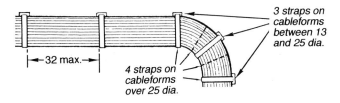

3 straps on cableforms between 13 and 25 dia.

32 max.

4 straps on cableforms over 25 dia.

NOTE: When using nylon Tyrap the distance between straps on straight parts of the cableform with no branches must not exceed 32 mm on cables up to 25 mm diameter. For over 25 mm diameter spacing may be increased to average diameter × 75 mm.

Fig. 1.8 Main loom strapping.

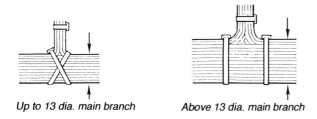

Up to 13 dia. main branch   *Above 13 dia. main branch*

Fig. 1.9 'T' Junction strapping.

NOTE: At 'T' branches it is permissible to use one strap applied in a 'figure of eight' for cableform up to 13 mm diameter but if the cableform exceeds 13 mm diameter use three straps.

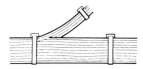

Fig. 1.10 'Y' Junction strapping.

NOTE: At 'Y' junctions a strap must always be applied to the main branch immediately before the junction.

Spot ties

Each spot tie must be a clove hitch with an overhand knot added.

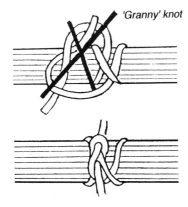

'Granny' knot

Fig. 1.11.

The tie must be neither too tight nor too loose. It must not:

■ Pinch the cable form

Pinching cableform

Fig. 1.12.

■ Slip along the cableform under gentle pressure

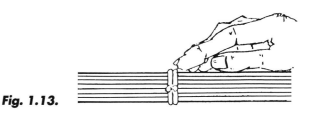

Fig. 1.13.

Check the spacing of the ties.

On cableforms of less than 25 mm diameter they should usually be 12 mm to 18 mm. On larger cableforms this distance should be roughly the diameter of the wire, unless otherwise specified. The spacing is closer on bends.

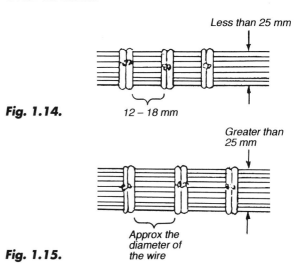

Less than 25 mm

Fig. 1.14.   *12 – 18 mm*

Greater than 25 mm

Fig. 1.15.   *Approx the diameter of the wire*

Check starting ties.

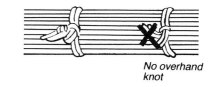

No overhand knot

Fig. 1.16.

Check finishing ties.

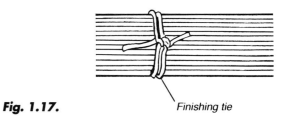

Fig. 1.17.   *Finishing tie*

The correct locking stitch must be used for the lacing. For cableforms of greater than 25 mm diameter a double locking stitch should be used.

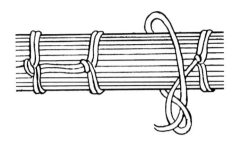

Fig. 1.18.

The tightness and spacing of the stitches should be as for spot ties.

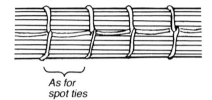

As for
spot ties

Fig. 1.19.

The lacing must not float round the bends.

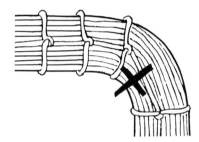

Fig. 1.20.

The stitches should be closer together to avoid this.

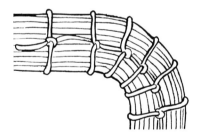

Fig. 1.21.

Ensure that branches and spurs are properly reinforced.

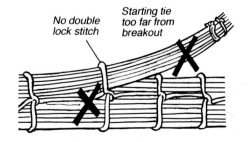

No double
lock stitch

Starting tie
too far from
breakout

Fig. 1.22.

Multistrand breakouts should have a double lock stitch before the branch. The starting tie on the branch should be close to the start of the branch.

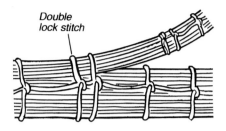

Double
lock stitch

Fig. 1.23.

Spurs should be reinforced by double locking stitches on both sides of the spur.

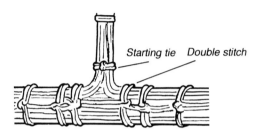

Starting tie   Double stitch

Fig. 1.24.

Sleeving

Sleeves can be used to bind wires together and to prevent short-circuits between electrical joints.

Sometimes, a heat shrinking sleeve may be used. Ensure that this is the correct diameter and that it is correctly applied. The hot air gun must be applied only to the sleeve. Care should be taken not to damage other wires or sleeving.

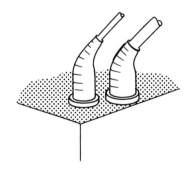

Fig. 1.25.

Straps

Check that each strap has been secured correctly, and lacquered with adhesive where required. Straps must meet the same conditions for spacing and tension as spot ties and lacing.

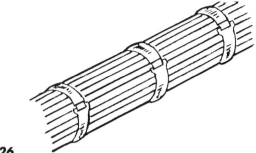

Fig. 1.26.

F – Twisted pairs

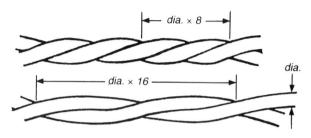

Fig. 1.27.

NOTES:

(1) The linear distance occupied by one 360° twist will be proportional to the outside diameter of the insulated wires and this distance must be not less than 8 times and not more than 16 times the overall diameter of one of the insulated wires and reasonably constant throughout the full length of the twisted pair.

(2) The insulations of the two wires should be in contact with each other throughout the full length of the twisted pair and only branch apart at the terminating points. This condition must exist if the wires are single-core conductors. If the wires are multi-stranded then it is acceptable to allow small spaces between the two wires providing that such spaces are not larger than 0.2 times the overall diameter of the wire.

(3) The pair should be bound together at the termination point with a rubber sleeve, nylon or PVC tie to prevent unravelling of the twist, except in the case of multiple twisted pairs where unravelling is not possible.

Wire identification

Colour coding is used to identify individual wires in equipment.

Colour code

Each colour refers to a particular number.

Black	0	Green	5
Brown	1	Blue	6
Red	2	Violet	7
Orange	3	Grey	8
Yellow	4	White	9

Coloured insulation

There are only ten alternatives. Each colour corresponds to one of the ten colour code numbers.

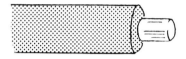

Fig. 1.28 Coloured insulation.

Multicoloured insulation

A large number of combinations can be obtained using coloured stripes on white wire. For three stripes the code is read by starting with the colour of the *lowest* number and moving along the wire in the direction of increasing colour value, e.g. black–red–green. For two stripes the missing third stripe is considered to be White (9).

Fig. 1.29 Multicoloured insulation.

Coloured markers

These are not often used. More frequently such markers carry a printed number. The sleeves are placed at each end of the wire.

Part number markers

On specified cables, sleeves carrying part numbers are fitted.

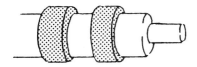

Fig. 1.30 Coloured markers.

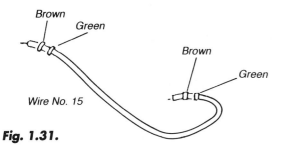

Fig. 1.31.

Typical cable

Single-core single-strand

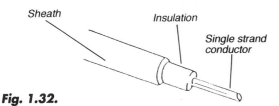

Fig. 1.32.

Three-core multistrand

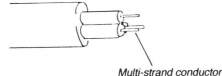

Fig. 1.33.

Screened single-core cable with no sheath – not often used.

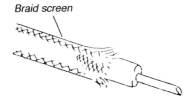

Fig. 1.34.

Screened two-core cable

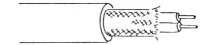

Fig. 1.35.

Coaxial cable is similar to screened cable except that the inner insulation is made of a special material and is dielectric. The sheath is usually black or blue in colour. This is the most commonly used screened cable.

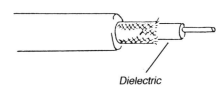

Fig. 1.36.

Conductor ratings

SWG 22 = 1–3 A
SWG 20 = 3–4 A
SWG 18 = 6–8 A
SWG 16 = 9–12 A

Sleeve stretcher

A typical sleeve stretcher has three spikes which move apart as the handles are pressed.

(1)   A sleeve of size somewhat smaller than the diameter of the wire is used

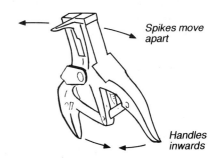

Fig. 1.37.

(2)   Place the sleeve over the spikes

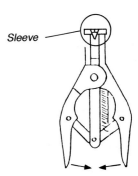

Fig. 1.38.

(3) Press the handles to stretch the sleeves to a size greater than the diameter of the wires. Do not overstretch

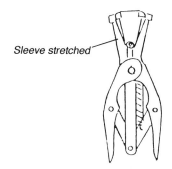

Fig. 1.39.

(4) Position the sleeve stretcher so that the sleeve is in the required position around the wires
(5) Ease the pressure on the handles while gripping the sleeve and pull the sleeve stretcher away

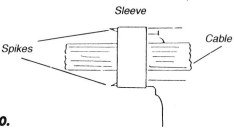

Fig. 1.40.

Heat shrinking sleeves

NOTE: Heat shrinking of sleeves over electrical joints is only carried out after inspection of the joint.

■ A special heat shrinking sleeve just larger than the diameter of the wire is used
■ Place the sleeve over the wires
■ Heat the sleeve with a hot air gun. Play the hot air evenly over the surface of the sleeve until it has shrunk to a snug fit

SAFETY: The hot air can be dangerous to both the operator and the equipment. Direct it only at the sleeve.

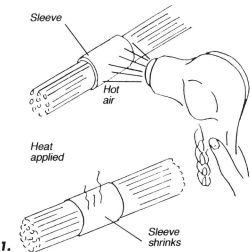

Fig. 1.41.

Tyraps or tie straps

Tie straps can be used instead of lacing in certain types of work. They are also used for mounting certain components.

Adjustment of tool tension

Fully squeeze trigger in order to apply required tension on the tool. Adjust the indicator in the window according to required stage by means of adjusting screw.

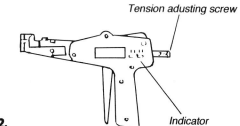

Fig. 1.42.

Operation

(1) Pass end of tie around cable loom or component and back through closure head

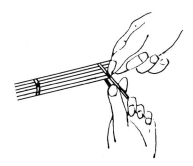

Fig. 1.43.

(2)   Pull up hard with fingers

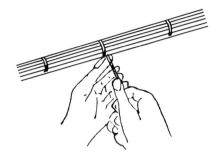

Fig. 1.44.

(3)   Hold firm the end of tie by hand and place tool inclined on the tie. Guide tool on tie, in inclined position, towards the cable loom until the tie head bears on the tool head

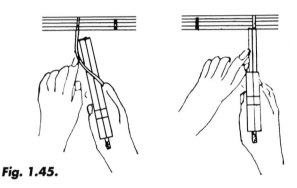

Fig. 1.45.

(4)   The cable tie is tightened by pulling the trigger of the cable tie gun once, or if necessary, several times. If the preselected tension has been attained, the cable tie will be cut off after the trigger has been pulled all the way to its stop.
(5)   Tyraps should be positioned exactly as indicated on the drawing, they should be tight and the tails should be cropped off level with the body.

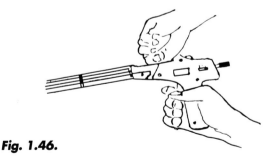

Fig. 1.46.

Forming pigtails – method 'A'

Form the braid into a pigtail to make an electrical connection.

Using a plunging tool

(1)   Remove the outer insulation
(2)   Push the braid back slightly to loosen it
(3)   Push the plunging tool under the braid until the tip eases between the braids

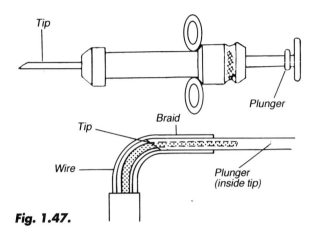

Fig. 1.47.

(4)   Press the plunger in gently. This will ease the wire from the braid
(5)   Straighten out the braid and sleeve
(6)   Form inner parallel to braid with no severe bends

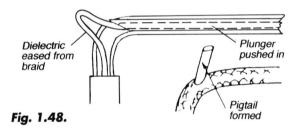

Fig. 1.48.

Forming pigtails – method 'B'

(1)   Prepare as shown, ensuring braid is neatly trimmed

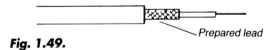

Fig. 1.49.

(2)   Carefully solder black sleeved fly wire. Do not damage dielectric insulation

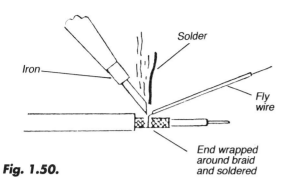

Fig. 1.50.

(3)   Finish as shown

Fig. 1.51.

Sleeving

Sleeves are used to bind wires together and to prevent short-circuits between electrical joints.

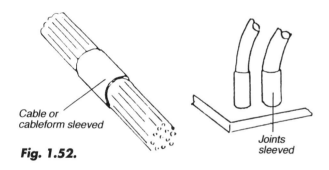

Fig. 1.52.

Twisted pairs

A wiring diagram may indicate that a twisted pair is to be used to connect terminals together.

Fig. 1.53.

To make a twisted pair:

(1)   Double over a piece of wire

(2)   Place the two ends in a hand drill chuck

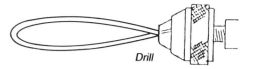

(3)   Hold the other end securely in a vice

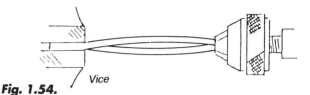

Fig. 1.54.

(4)   Rotate the hand drill so that a twisted pair is formed

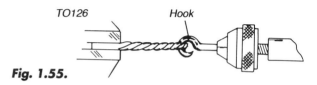

Fig. 1.55.

Where leads must be twisted together ensure that they are:

- Twisted evenly
- Twisted neatly without kinks or loops
- Twisted tightly enough to prevent loops from appearing easily
- Not twisted so tightly that stress is caused on the wires or joints
- Where wires are shown straight they should be straight, i.e. twists to be removed

NOTE: Drawings should show the number of twists per centimetre or inch, the presence of any additions such as ferrite beads or spacers and whether the twisting is even or in a particular area.

7/0.2 16/0.2	between 3 & 4 twists per inch (25mm)
30/0 63/0.2	2 twists per inch (25mm)
Twisted triple 16/0.2	approx: 1 per inch

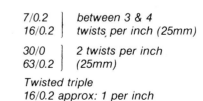
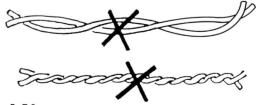

Fig. 1.56.

A twisted pair can be made by hand but care must be taken to twist evenly and to avoid kinks and loops.

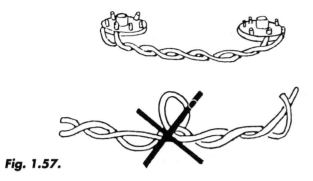

Fig. 1.57.

General

Cableforms are inspected as a subassembly before they are fitted into the equipment. The inspection of a completed cableform involves:

(1)   The inspection of each individual wire.

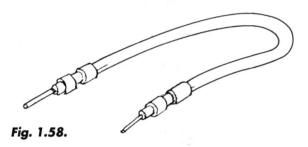

Fig. 1.58.

(2)   The visual inspection of the completed cable-form.

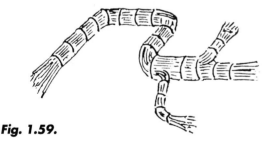

Fig. 1.59.

(3)   Inspecting the assembly and wiring of plug and socket connectors.

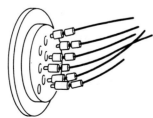

Fig. 1.60.

(4)   The electrical testing of the completed cable-form.

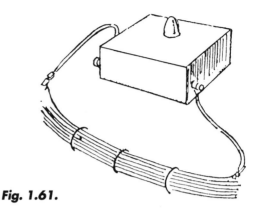

Fig. 1.61.

Potting

Where cables are attached to plugs or sockets and the joints are potted to provide added security, the cables should be inspected both visually and electrically before and after potting to ensure that the potting process is not damaging the joints.

Individual wires

Inspect each wire against the wiring schedule. Check for correct length, type and gauge.

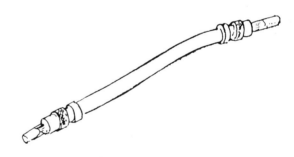

Fig. 1.62.

Stripping and tinning

Ensure that the correct length of insulation has been stripped from each end of the wire.

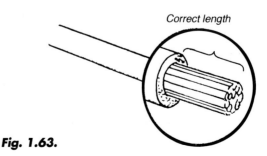

Correct length

Fig. 1.63.

Check for damage to the wire and insulation.

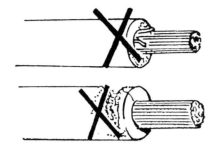

Fig. 1.64.

Pigtails and stranded conductors should be adequately twisted. There must be no loose, broken or doubled up strands.

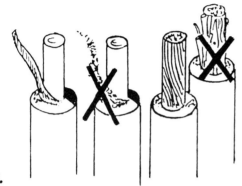

Fig. 1.65.

Routing of wires

Route all wires, cables and cableforms away from contact with sharp objects, moveable parts and heat generating components.

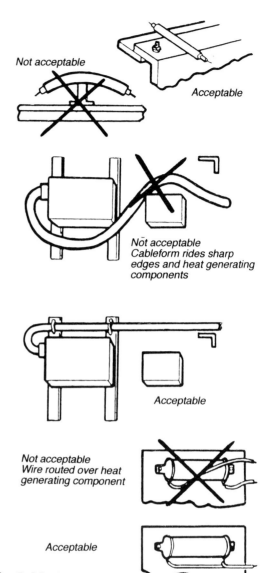

Fig. 1.66.

Make sure components and subassemblies can be easily removed without damaging surrounding components.

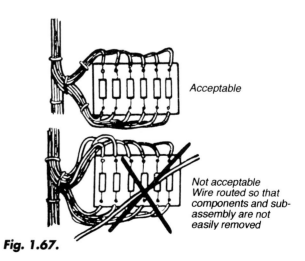

Fig. 1.67.

Use a suitable grommet when routing wires through holes in metal plates.

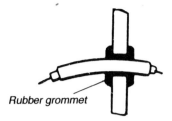

Rubber grommet

Fig. 1.68.

The cableform must be routed and secured so as to avoid:

■ Sharp edges such as metal corners or switch tags

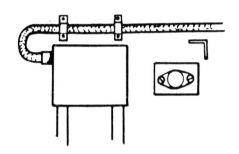

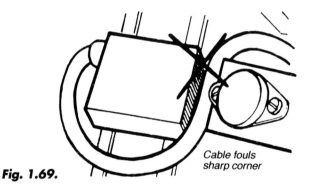

Cable fouls sharp corner

Fig. 1.69.

■ Heat generating components
■ Moving parts

A grommet or eyelet should be used wherever the cableform is passed through a hole in metal.

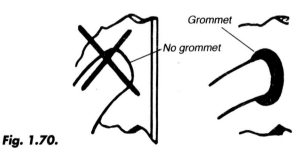

Grommet

No grommet

Fig. 1.70.

Where possible the cableform should be routed so that component identification is not unnecessarily obscured and so that removable parts or modules are not trapped by the cableform.

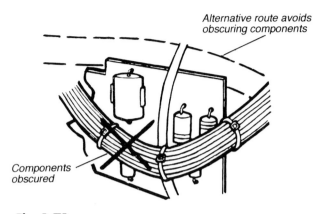

Alternative route avoids obscuring components

Components obscured

Fig. 1.71.

Screened leads

Ensure that the conductor and the braid are connected to the correct connecting point.

There should be greater stress relief on the conductor so that any stress is taken on the braid.

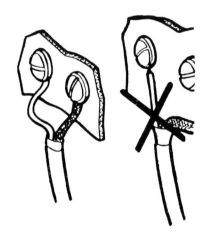

Fig. 1.72.

Electrical tests

Carry out any tests on the equipment as instructed using the specified continuity tester.

Carry out the test according to a point to point test schedule. This presents the information shown on the wiring diagram in a simpler form, and ensures that no checks are missed.

Table 1.1 Point-to-point test schedule.

From	To	Min. Ω	Max. Ω
Point A	Point B	0.00	0.04
Point C	Point D	10 M	∞

G Termination of wiring

1. TERMINATIONS

Termination of wiring in the context of this section is defined as any point in an electrical circuit where the material of the conductor becomes discontinuous and it is necessary to make a good electrical joint to continue the circuit, i.e. input terminal to connecting wire, connecting wire to component leadout, etc.

Circular cross-section conductor must always be designed to terminate with a soft soldered joint wherever this is possible (see Section 4). A soft soldered joint is the simplest and most reliable method of ensuring a maximum area metal to metal permanent contact.

Wire wrapping may be used as an alternative wire terminating method but only if the full specification requirements can be complied with.

Electrical connections must never be made by clamping round section wires between flat surfaces, such as two nuts on a threaded terminal. A flat solder tag, crimp tag or cable lug of the correct type must always be used under these circumstances.

The soldering joint between flat solder tags or cable lugs and connecting wires need not be protected by a rubber or synthetic sleeve unless the wire will be moved during operation or servicing of the equipment.

Soldered connections between connecting wires and plugs and sockets must always be protected by a rubber or synthetic sleeve.

Electrical connections must never be made by clamping between metallic and insulating materials as most insulating materials are subject to shrinkage which will relax the clamping pressure.

If a hank bush or spacer is riveted to a PCB and is required to form part of an electrical circuit between the track of the PCB and the chassis on which the board is to be mounted, then the hank bush or spacer must be soldered to the PCB track for at least one-third of its circumference.

When wires are required to be terminated at screw type terminals the following methods are to be used:

- For plain nuts or screw-down type terminals or terminal strips of the 'cinch' type where the connection is made under the nut or the head of a screw, a 'flag tag' or 'spade' type terminal should be fitted to the wire, preferably of the type which can be clenched or crimped on the insulation of the wire. For 'Cinch' type terminal strips a fanning strip should normally be used.

- For 'Grelco' type connecting blocks (the wire is secured in a hole in a metal block by radial pressure from the end of a screw pressing directly on the wire). The bare end of the wire must be bent over double on itself and the wire inserted into the hole in the block so that the insulation touches the metal block before the screw is tightened. If the wire is stranded, the strands must be laid and twisted up neatly before bending but the wire should not be tinned. The clamp screw must not be overtightened or the wire strands will be cut. For mobile installations or any connection which will be subject to a large amount of vibration this termination must not be used but a cable thimble of the 'crimped on' type which also grips the insulation of the wire should be fitted.

- For the termination of the 'Klipon' block type (wire is clamped between two flat metal plates by the action of a screw) either the correct crimp on wire termination must be used or the bare wire must be bent back on itself before insertion between the clamp plates. If the wire is stranded it must not be tinned.

In installations not subject to excessive vibration, it is permissible to connect more than one wire to one connecting block of the 'Grelco' or 'Klipon' type. In this case the individual wires need not be bent back upon themselves. They must not be twisted together or tinned if stranded. The number of wires connected onto one block must be limited to the number which can be accommodated in the block with ease with a maximum number of four. When more than one wire is fitted into a single 'Klipon' type block then all the wires must be of the same type and size.

Care must be taken when terminating wiring at rotary wafer switches to ensure that the solder tags (which are the outer ends of the contacts) are not put under strain while soldering or afterwards due to the attachment or dressing of the wiring. The following points should be observed:

- The set of the tags must not be altered (tags must not be bent) unless this is absolutely necessary and then only when specified in the drawings or assembly instructions. Great care must be taken not to loosen the securing rivets as the tag position is being reset

- Where possible, connecting wires should approach the switch in an easy semicircular sweep and in one direction only on each wafer, in any case the tag must never be under tension due to being pulled by the wiring

- The solder joints must conform to the manufacturers' requirements or the requirements in Section 4. The minimum of solder must be used and the joint made as quickly as possible, consistent with the time required to make a good joint, to ensure that the solder does not run down the tag towards the contact surfaces or overheat and loosen the contact assembly

- If it is necessary to link two or more tags together on the same switch wafer then this must be done using bare tinned copper (BTC) wire not larger than 0.560 mm (24 SWG). If the linked tags are not adjacent then the wire must be sleeved to pass the nonconnected tags and must not rest against these tags

- Before attempting to clean rotary wafer switches in any way, reference must be made to Section 5

It is permissible to use crimped electrical connections for the termination of wiring on manufactured equipment providing that:

- The method is approved by the design authority for the equipment concerned and crimped electrical connections are specified in the drawings

- The crimping methods used satisfy the requirements of the relevant British Standard

H Wire-wrapped joints

1. SCOPE

This standard establishes the requirements to produce mechanically and electrically stable, solderless wrapped, electrical connections made with single, solid round wire and appropriately designed terminals.

This standard includes classes and requirements for solderless wrapped connections, the visual inspection, mechanical and electrical testing.

2. CLASSIFICATION

This standard covers the following classes of solderless wrapped electrical connections:

Class A   modified solderless wrapped electrical connections
Class B   conventional solderless wrapped electrical connections

3. CLASSES OF SERVICE

The classes of service, based on the vibration environment are:

Class A   severe
Class B   moderate

Class A – severe

Severe vibration is defined as the vibration encountered by mobile equipment, equipment repeatedly transported during its life, or by equipment subjected to other environments exhibiting appreciable vibration such as ship board, airborne or space applications. Modified solderless wrapped connections should be used for Class A service (see Fig. 1.73).

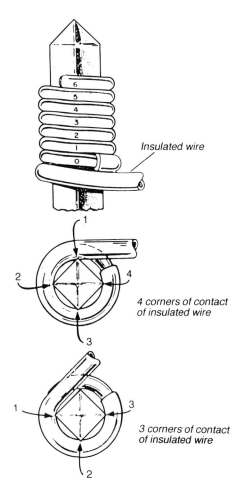

Fig. 1.73.

Class B – moderate

Moderate vibration is defined as vibration no worse than the negligible vibration encountered by stationary equipment at the operational site and the vibration during infrequent shipment of the stationary equipment.

Conventional solderless wrapped connections may be used for Class B service (see Fig. 1.74).

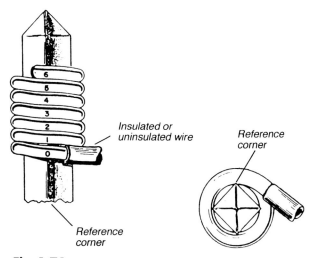

Fig. 1.74.

4. DEFINITIONS

The following is a list of key terms used in this standard.

Solderless wrapped electrical connections

The terminology and make-up of this type of connection on one type of wrap-post configuration is illustrated in Figs 1.73 and 1.74.

Class A (modified solderless, wrapped connection)

This connection consists of a helix of continuous, solid, uninsulated wire tightly wrapped around the wrap-post of a solderless wrapped contact to produce a mechanically and electrically stable connection. The number of turns required will depend on the gauge of wire used. In addition to the length of uninsulated wire wrapped around the wrap-post, an additional minimum half turn of insulated wire shall be wrapped around the wrap-post to help ensure better vibration characteristics. To accomplish a half-turn, the wire must be in contact with at least three corners of the wrap-post as shown in Fig. 1.73.

Class B (conventional solderless wrapped connection)

This connection is the same as described in Fig. 1.73 except that the additional half-turn of insulated wire is not required as shown in Fig. 1.74.

End tail

An end tail is the end of the last turn of wire of a solderless wrapped connection which may extend in a tangential direction instead of resting against the wrap-post (see Fig. 1.75).

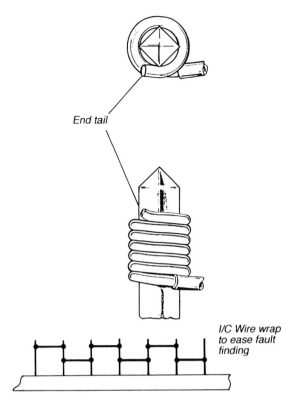

End tail

I/C Wire wrap
to ease fault
finding

Fig. 1.75.

A turn of wire

A turn of wire shall consist of one complete, single helical ring of wire wrapped 360° around a wrap-post. A connection having N turns in contact with the wrap-post will intersect the reference corner $N+1$ times.

Gas-tight area

The gas-tight area is that contact area between the wrap-post and wire which, due to the quality of the wrap, will exclude gas fumes.

5. GENERAL REQUIREMENTS

Description

Solderless wrapped connections shall be made by wrapping a specified number of turns of wire, under tension, around a post having sharp corners. The sharp corners of the wrap-post shall produce high pressure points resulting in indentation of the wire or both the wire and wrap-post to meet the requirements of the strip test.

The resulting gas-tight high pressure points shall provide electrical continuity and mechanical stability.

Connection parameters (visual requirements)

■ There shall be no gaps between adjacent turns greater than one-half of the diameter of the wire exclusive of gaps on the first and last turns
■ The sum of all gaps (excluding those on the first and last turns) on any side of a connection shall not exceed the diameter of the wire
■ There shall be no overlapping within the minimum specified number of turns of uninsulated wire
■ It is not permissible to rewrap the portion of wire that has been previously wrapped on a terminal

6. MINIMUM NUMBER OF TURNS

There shall be a sufficient number of turns of uninsulated wire in the connection to ensure conformance with all the visual, mechanical and electrical requirements of this standard.

Table 1.2 Minimum number of uninsulated turns of wire.

SWG	Cu wire	Cu-alloy	Ni-Fe/Cu clad steel
30	7	5	–
29	7	–	–
28	7	–	–
26	6	–	5
25	5	–	5
24	5	–	4
22	4	–	4
20	4	–	4
18	4	–	3
16	3	–	3

7. MARRING & SCRAPING

Mars and scrapes produced on the wire in the connection, including the connector lead as a result of wrapping, or as a result of stripping (skinning) are permissible provided that the connection meets the requirement of this standard.

8. SOLDERLESS WRAPPED CONNECTION ON THE SAME TERMINAL WITH SOLDERED CONNECTION

Solderless wrapped connections and soldered connections may be used on the same terminal provided that the terminal length for the solderless wrapped connection is free of excess finish coat deposited from the soldering process. The order of placement of solderless wrapped or soldered connections on the terminal is not important provided the requirements of paragraph 10 are met.

9. MECHANICAL TESTS

Strip force

A completed solderless wrapped connection shall be capable of meeting the following minimum strip force limits. The minimum number of turns used on the product shall not be less than the minimum number used for strip force testing.

Table 1.3 Minimum strip force.

Wire SWG	Minimum strip force		
	lb	kg	N
30	3.0	1.36	13.6
29	3.5	1.59	15.9
28	4.0	1.81	1.81
26	5.0	2.27	22.7
25	6.5	2.95	2.95
24	6.5	2.95	2.95
22	8.0	3.63	3.63
20	8.0	3.63	3.63
18	12.0	5.44	54.4
16	15.0	6.80	68.0

Strip force test

The strip force is the force required to cause the initial breakaway of the connection.

The strip force test shall be performed as shown in Fig. 1.76 and the measurement shall meet the requirements of Table 1.3.

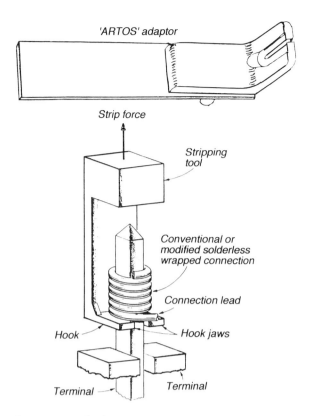

Fig. 1.76 Strip force test.

Stripping tool

The Artos pull-off gauge shall be used, the jaws of the stripping tool hook shall be vertical to the longitudinal axis of the terminal, creating a flat surface contact with the wire on either side of the terminal at the lead end of the connection.

The jaws of the hook shall engage along the major dimension (cross-section) of the terminal. The maximum total clearance between the jaws and terminal shall not exceed $0.7 \times$ the diameter of the wire. The minimum clearance, when the terminal and stripping tool are properly aligned, shall be such that there is no binding between jaw and terminal.

The stripping rate should be 240 mm per minute and the direction shall be towards the free end of the terminal, the strip force being applied along the axis of the terminal.

Unwrap test

The wire of a connection should be capable of being sufficiently unwrapped to free the wire from the terminal without breaking.

Unwrap test (conductor embrittlement)

Place the unwrapping tool over the terminal post and engage its leading edge between the wrap end and the next wrap turn. Rotate the unwrapping tool until all the wire has been transferred onto the tool. Remove the tool with the loose helicoil from the terminal post. Holding the insulated portion of the wire firmly, rotate the tool unwinding the wire. The unwrapped wire need not be perfectly straight, waves and permanent deformation in it are permissible.

Failure of the unwrap test shall not be cause for rejection of the product, but shall serve as an indication for the necessity to examine the wrapping process. Some of the variables are:

- Faulty operator's technique
- Improper tool performance
- Detrimental wire and terminal characteristics

10. HANDLING PRECAUTIONS

Completed solderless wrapped connections shall not be mechanically disturbed. Mechanical disturbance is defined as any axial movement of the connection on the terminal or change in the configuration or appearance of the connection.

The wire shall be positioned so that subsequent routing of the lead end of the wire does not tend to loosen the connection.

11. ELECTRICAL TEST

When measured as in Fig. 1.77 at the current specified in Table 1.4, the voltage drop across the wrapped connection shall not exceed 4 mV.

Table 1.4 Test current.

SWG	Current used to determine wrapper resistence (A)
30	1.0
28	2.0
26	2.4
24	2.4
22	2.4
20	7.5
18	7.5

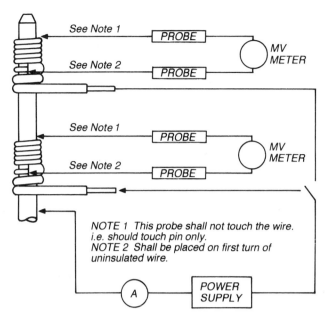

NOTE 1 This probe shall not touch the wire.
i.e. should touch pin only.
NOTE 2 Shall be placed on first turn of
uninsulated wire.

Fig. 1.77.

12. INSPECTION OF TOOLS & SOLDERLESS WRAPPED JOINTS

Wrapping tools

Solderless wrapped connections are formed by special tools which are available in hand and powered versions. The design requirement is that the tool shall produce consistent connections in respect of the number and spacing of turns and neat closure of the last turn, without causing damage to the wrapping wire or terminal. Tools are evaluated by testing sample connections for quality and consistency.

Unwrapping tools

Unwrapping tools must not damage the terminals and must positively remove the unwrapped wire from the equipment.

Tool control

- All tools must bear individual serial numbers
- Tools in use shall be inspected daily for damaged working faces
- Any tool which has been dropped or severely jarred shall not be used again until a member of the quality control staff has checked that its performance is still satisfactory
- A record of connection sampling test shall be maintained for each tool

■ For each test operator combination per week, a set of six joints shall be made

■ A strip force test shall be applied to three joints of each set and shall be carried out as described in subsection 9 of this Section and using the force specified in Table 1.3. The remaining three joints of each set shall be unwrapped, using a rotary motion in the plane of the helical turns, avoiding any twisting action or radial tension. It must be possible to unwrap each joint without breaking the wire

Inspection of joints

Every joint shall be visually examined to ensure that the correct combination of tool, wire and terminal has been used, and for compliance with the following detailed requirements:

■ The connection shall have a minimum number of turns as specified in Section 6. When covered wire is used, the class of connection will be Class B unless the specification states otherwise

■ The turns of a connection shall be close wound without overlap, and comply with this standard

■ Where wiring is to be 'dressed' prior to wrapping, sufficient slackness shall be allowed to ensure that no strain is placed on the connection and that sufficient wire is allowed for rewrapping if necessary

■ Connections made on a common terminal shall not touch or overlap. No more than three wrapped connections may be made to any one terminal

■ The preferred direction of wrapping is clockwise

■ A joint once made shall not be disturbed and there shall be no attempt to tidy up the end or tighten the joint

J – Insulation stripping

General

When insulation is removed from a wire prior to making a joint the utmost care must be taken to see that the metal conductor (or braid in the case of a screened or coaxial wire) is not nicked or damaged in any way by the stripping operation. The following

paragraphs describe the methods and tools to be used.

Plastic insulation

(PVC, polythene, etc.)

■ Correctly adjusted hand operated wire stripper
■ Hot 'vee' stripper
■ Rush rotary brush stripper

Enamelled wire

■ Emery paper not coarser than 1M grade
■ Burn in small naked flame (alcohol burner or gas) or hot solder pot and clean off charred residue with emery paper not coarser than 1M grade
NOTE: Cleaned wire must be correctly tinned as soon as possible after the cleaning operation
■ Rush rotary brush stripper
■ Polyurethane enamel insulation (Bicelflux) removed by soldering operation. This requires a longer application of heat than a similar sized joint made with pre-tinned wire to be certain that the insulation has melted and the wire has tinned correctly

The standard length of strip for single wires (solid or stranded conductor) is 20 mm.

The stripped wire length to be adjusted during making of the joint so that the bare conductor at the joint is not greater than 1.5 mm.

K – Damage to wires

Insulation stripping should not damage the wire. In the case of multistrand wires, all strands should be present and unmarked.

Defect classification

Any missing strands or nicks up to 10% of the wire diameter are a Class C defect. More than 10% of strands missing or more than 10% of damage is a Class B defect.

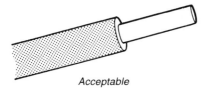

Acceptable

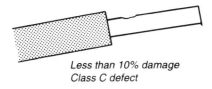

*Less than 10% damage
Class C defect*

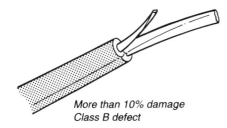

*More than 10% damage
Class B defect*

Fig. 1.78 Damaged stripped wire ends.

M – Mechanical stress on joints

Wires should be ranked and tied off in such a way as to ensure that there is no mechanical stress on the joint.

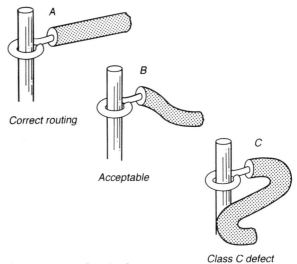

Correct routing

Acceptable

Class C defect

Fig. 1.80 Mechanical stress.

L – Excess or insufficient insulation clearance

Incorrect clearance of insulation from the joint may cause either dry joints, or risks of shorts.

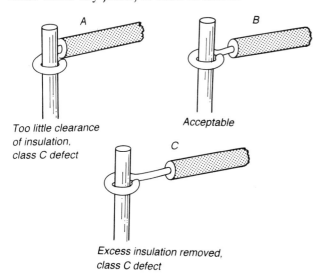

*Too little clearance
of insulation,
class C defect*

Acceptable

*Excess insulation removed,
class C defect*

Fig. 1.79 Insulation clearance.

N – Turns around pins

A wire should be terminated between a minimum of half a turn (180°) and two turns maximum as illustrated below.

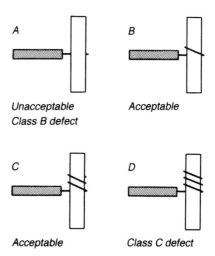

*Unacceptable
Class B defect*

Acceptable

Acceptable

Class C defect

Fig. 1.81 Turns around pins.

P – Turns around tags

A wire should be terminated between a maximum of half a turn (180°) and two turns maximum as illustrated below.

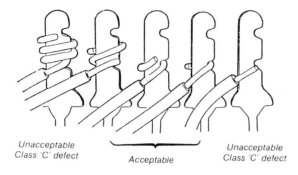

Unacceptable Class 'C' defect   Acceptable   Unacceptable Class 'C' defect

Fig. 1.82 Turns around tags.

Q – Connection to tags

All joints must be completely separate terminations.

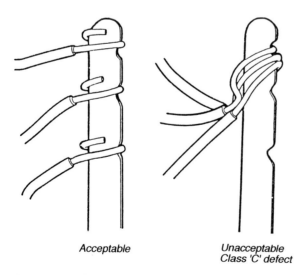

Acceptable   Unacceptable Class 'C' defect

Fig. 183 Multiple connections to tags.

NOTE: The same connection principles apply to multiple connections to pins.

R – Crimp connections

1. INTRODUCTION

The method of crimping wires into connectors is a widely used type of termination whereby the connector and the wire are joined together by a crimping method. The main advantage of using this method is that there is no need to destroy the connection when fault finding, i.e. most crimp tags and connectors can be mechanically removed without damaging the terminations. The other distinct advantage is that repairs can be carried out without the need for electrical power as the standard hand tool can consistently accomplish satisfactory connection using semiskilled labour.

2. BASIC CRIMP CONNECTIONS

Faston connector

Closed barrel types

Open barrel types

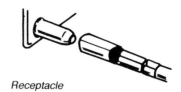

Receptacle

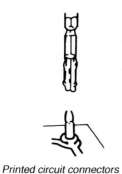

Printed circuit connectors

Fig. 1.84 Basic crimp connections.

3. CRIMP INSPECTION

Visual

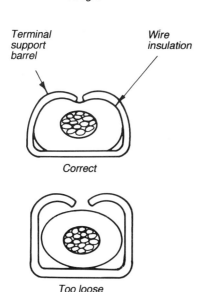

Too tight

Terminal
support
barrel

Wire
insulation

Correct

Too loose

Fig. 1.85 Receptacle.

NOTE: The visual appearance of the crimp tag is a good indication of the quality of the joint.

Manual physical check

It is not necessary to exert excessive force whilst carrying out this check. If a gap appears between the tag and the insulation showing the conductor, or if the wire snaps or pulls completely out of the tag then the crimp is faulty.

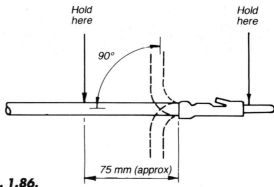

Hold here   Hold here

90°

75 mm (approx)

Fig. 1.86.

Should the sample of tags inspected show any failures, the tools should be checked for wear and tear, correct jaws/setting, etc. The operator should also be informed of the problem and reinstructed in the use of the crimp tools as necessary.

Mechanical pull-off test

A mechanical pull-off test will enable the measurement of definitive pull-off force. This will give a more accurate indication of the mechanical quality of the crimp joint. The method of applying the test is simple and the test jig will consist of the following:

- A calibrated maximum indication gauge either a scaled mechanical indicator or a calibration dialled instrument
- A method of holding the crimp tag
- A method of applying mechanical stress on the joint, e.g. by applying weights in a simple pulley system. Alternatively the pull-off force can be applied hydraulically (special purpose-built test rigs are available) or by adapting a pillar drill to apply the force (see Fig 1.87)

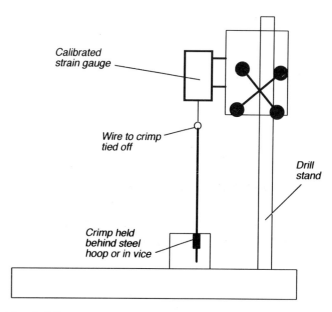

Calibrated strain gauge

Wire to crimp tied off

Drill stand

Crimp held behind steel hoop or in vice

Fig. 1.87.

Strain on the crimp is increased:

(a)   The wire pulls out of the crimp;
(b)   The maximum stipulated force has been applied.

Each crimp tag manufacturer recommends the minimum pull-off test for each type of crimp tag and it is normal to take a minimum of three crimp joints per batch and the average pull-off force must meet these minimum requirements.

A record of all test crimps against crimp tools shall be kept, normally by placing a tie-in tag on the crimp with the serial or identification number of the tool together with the time and date.

4. CRIMP TOOLS

There are various types of crimp tools ranging from the simple hand tool, single action through to the hand ratchet type to the hydraulic semi-automatic and finally to the fully automatic type.

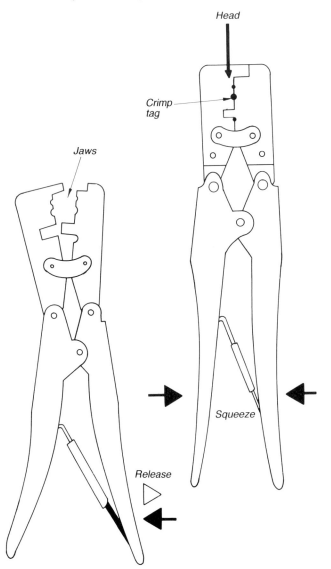

Fig. 1.88.

Basic crimp tool

This type of crimp tool is simple to operate and relies on the strength of the operator to give a good crimp joint. NOTE: heads need to be changed in order to cover the full range of cable and corresponding crimp sizes used in the electronic industry.

Ratchet crimp tools

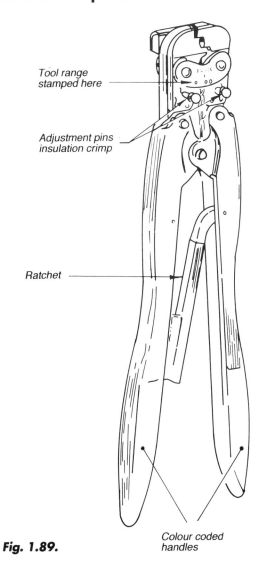

Fig. 1.89.

Points to note

The ratchet crimp tools are normally set up or calibrated by the tool supplier and in order to avoid the wrong tools being used on the wrong crimp tags, the tags and the tool handles are colour coded to correspond to each other:

Red tag   Red handles
Blue tag   Blue handles
Yellow tag   Yellow handles
and so on

NOTE: It is important that each production area displays a table or chart which sets out the colour of the handles against their corresponding crimp tags.

Inspection points to note

When checking a crimp joint the following general points are relevant to all types of crimp tag.

- The wire must be visible at either end of the wire barrel or wire contact area
- The insulation must be visible beyond the insulation barrel. NOTE: The wire and the insulation should be visible in the area between the insulation and the wire contact areas
- The connector must be free from fractures (see Fig. 1.90)
- There must be the correct gap between the insulation and the crimp tag (see Fig. 1.91)
- There must be no loose strands of wire outside the crimp tags (see Fig 1.92)
- The insulation must be held firmly without causing damage (see Fig. 1.93)
- Screen cable must be crimped such that the screen is included with the crimp area (see Fig. 1.94)

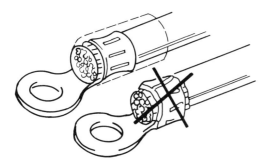

Fig. 1.90 Connector fractures.

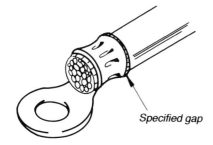

Specified gap

Fig. 1.91 Insulation gaps.

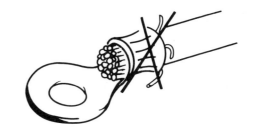

Fig. 1.92 Loose strands.

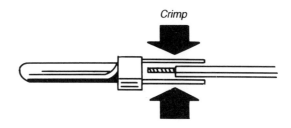

Crimp

Fig. 1.93 Insulation crimp.

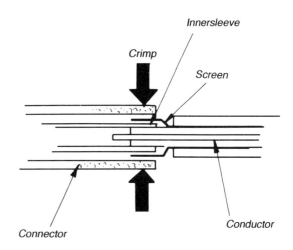

Innersleeve

Crimp

Screen

Connector

Conductor

Fig. 1.94 Screen cable.

5. CRIMP TOOLS CHECKS

Carrying out crimp pull-off tests checks both the operator and the crimp tool for technical performance. The appearance of the crimp joint is also important and can give an indication of possible problems before they arise (see Fig. 1.85).

NOTE: A record should be kept showing the number of faults, the corrective actions taken and adjustments or repairs carried out on each crimp tool, and dates on which sample inspections were carried out.

S – Coaxial terminations

1. MALE CONNECTOR

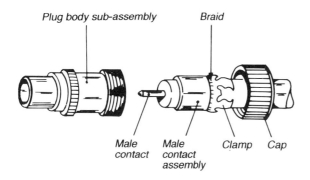

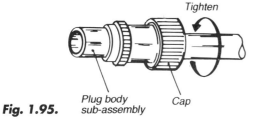

Fig. 1.95.

- No damage to outer insulation
- Assembly sequence is as per Fig. 1.95
- Clamp correctly fitted with claws grasping the insulation
- Braids neatly trimmed around body of clamp
- Conductor correctly soldered into male contact with the conductor wire flush, yet its outline visible, in the end of the contact
- Cap firmly screwed into position

2. FEMALE CONNECTOR

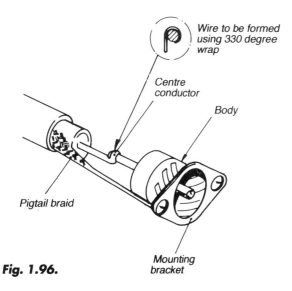

Fig. 1.96.

- The pigtail braid can either be soldered to the body of the connector or alternatively, soldered to a tag which is in turn secured between the socket fixing screws and the equipment chassis/panel. It is important to note that the securing point on the chassis/panel must be free of all paint or other insulating treatments to ensure correct electrical 'binding' between the tag and the chassis/panel.

- The pigtail is normally sleeved to avoid accidental contact with the centre conductor

- Ensure there are no loose strands that can cause a short-circuit with the centre conductor

- It is common practice in mechanical vibration applications to encase the joint by means of a heat shrink sleeve or boot which provides mechanical strength to the joints as well as adding electrical insulation to the joints

T – Ribbon cables

1. CONDUCTOR COLOUR IDENTIFICATION

Conductor Number	Colour
1	Black
2	Brown
3	Red
4	Orange
5	Yellow
6	Green
7	Blue
8	Violet
9	Grey
10	White

Each subsequent block of ten conductors is colour coded the same, i.e.:

11	Black
12	Brown
13	Red

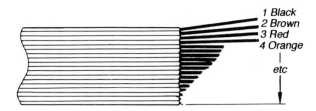

Fig. 1.97 Circuit sequence.

2. ASSEMBLY POINTS

The principle of operation in all ribbon cable assembly, no matter what type is used, is the same.

The outer visualisation of each individual conductor is pierced by a conductor pin and the cable firmly clamped in position so as to maintain electrical contact.

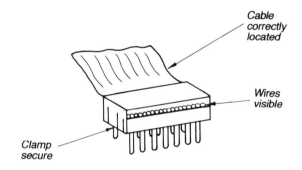

Fig. 1.98.

3. BASIC RIBBON CABLE TERMINATION

The connection shown is a basic dual in line (DIL) connector but the principle of ensuring contact with the ribbon cable is the same, no matter the type of connector used. The main points to note are:

The ribbon cable
- Cut square and cleanly
- Located correctly to the pins
- One conductor per pin
- Free of kinks, damage, etc.

Clamp
- Correctly located in relation to the cable and connector body
- Securely fitted

Body
- Connector pins free from damage
- Body not damaged

Assembled connector
- Wires visible from face connector

- Clamp securely located
- Clamp securely fitted
- Cable firmly clamped
- Continuity between conductors and their relevant conductors
- Contact pins undamaged (i.e. pins bent, damaged etc.)

NOTE: Always ensure that the correct tooling is used for the connector in hand.

U – Cubicle wiring

The term 'cubicle wiring' can be applied to any racking equipment where a combination of wiring is needed to accommodate mains supply, signal paths, interconnection between modules in the same cubicle, or rack of interconnecting between cubicles.

Several basic rules apply to cubicle wiring which, if consistently applied, will reduce the problems associated with this type of wiring. Some examples are:

- Interference between power and signal leads
- Mains-borne noise spikes affecting switching circuitry
- Radiation of noise causing local interference in the equipment rack (radio frequency (RF) interference)
- Radiation of noise causing interference in adjacent equipment
- Faulty earthing causing intermittent faults
- Faulty interconnecting continuity plug sockets, wiring etc

1. BASIC RULES

The following basic rules may not be fully applicable in all situations or, indeed, fully comprehensive in other situations. Suffice it to say that the following rules have been drawn up from experience gained in cubicle design and manufacture and must serve at least, as a check list for design and production engineers.

- Wherever possible, always route the power cables separately from (or to) the signal looms
- Always ensure that unnecessarily large loops of wire are avoided (only leave enough wire to remake the joint, should it be necessary)

- Where it is desirable to maintain separate signal earths and chassis earths they should be clearly identified. See the current IEE Regulation for the correct colour coding in this instance
  NOTE: It may be desirable for the signal earths and the chassis earths to be one and the same and the signal would normally be connected to the chassis earth via a resistor to enable monitoring to take place.
- All metal chassis and surfaces should be electrically bonded together to avoid earth loops, etc.
- All equipment looms should be correctly formed (see Section E)
- All wiring used for earthing should be a common colour, i.e. mains earth green/yellow
- Signal earths are normally black.
  NOTE: This may be subject to international variations.
- Screened leads directly connected to chassis earths should have their screen terminated by a green/yellow lead or 'pigtail'
- Screened leads having their screen terminated to signal earth would normally have the screen terminated via a black lead or pigtail'

- Where it is necessary to bond two moving surfaces, e.g. the door of a cubicle bonded to the main cabinet, it is necessary to 'spot face' an area on each surface to be bonded and joined by green/yellow wire terminated in tags bonded to the spot faces. Ensure that the green/yellow wire is sufficiently long to enable the door to be opened without disturbing the joints
- Each earthing point must be identified with the earth symbol (see Fig. 1.99)

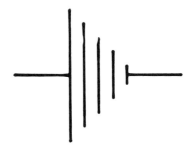

Fig. 1.99 Earth symbol.

SECTION 2: ASSEMBLY OF ELECTRICAL/ ELECTRONIC COMPONENTS

A – General

1. RULES AND PROVISOS

For convenience of manufacture electrical components used in any particular circuit are mounted together in groups on some form of insulating board or panel which may be secured into the equipment and which is supplied with connecting points or plugs and sockets to which connections may be made. The most notable exceptions to this general rule are large components such as mains transformers (which due to their physical size and weight must be mounted directly on a solid part of the chassis or frame) and variable components, switches, potentiometers etc., which are mounted on or behind the front panel to be used as equipment operating controls.

Terminals

Various types of terminals are used to make connections to PCBs. Terminals are assembled by:

■ Riveting. The terminal itself is a form of mast. The terminal is held by a serration or taper on the terminal
■ Screwing in
■ Staking

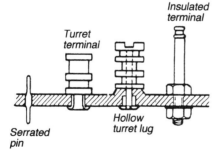

Fig. 2.1.

■ Ensure that the correct terminals have been used
■ Ensure that they are tight
■ Check for damage to the board or to the terminal, e.g. cracking around the base of the terminal is a common fault
■ Report all defects

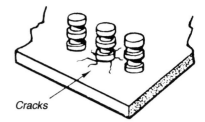

Fig. 2.2.

Small electrical components which are supplied fitted with moulded in connecting wires (resistors and capacitors) must not have such connecting wires extended by soldering a further length of wire to the existing wire lead. The full length of the component lead may be used for fitting the component into the circuit and supporting the body weight but the lead-out wires must always be soldered directly to a solder tag, spill, turret lug, stand-off terminal or PCB solder pad.

Small wire-ended resistors or capacitors may be soldered directly across the terminals of large components (e.g. large metal-cased capacitors) providing such terminals are designed to be used as solder anchor points. Normal joints must be made to solder tags fitted to the screwed terminals.

In certain existing designs transistors are fitted into the heat sink, clips or rubber grommets and long transistor leads (25–38 mm) are sleeved and soldered to spills or solder terminals mounted adjacent to the transistor. Certain of these transistors or their replacements are now only obtainable with short lead-out wires designed for PCB mounting. In these cases it is permissible to extend the lead-out wires and the following method will be adopted.

(1) A transistor mounting pad is to be placed over the straightened leads of the transistor and pushed down until it is in contact with the transistor body
(2) A closed ring is to be formed on the end of a piece of 0.560 mm (24 SWG) BTC wire which will slip easily over the straight lead-out wire of the transistor. A soldered joint is then to be made between the closed loop and the transistor lead-out wire while holding the wire loop pressed firmly down onto the mounting pad which is also to be pressed against the transistor body
(3) This procedure is repeated for each transistor lead-out. The wire extension must be cut not longer than 50 mm and should be sleeved with the appropriate red, green or white PVC sleeving. The transistor lead-out wires may be shortened but must project through the mounting pad by at least 3.5 mm and be cut before the joint is soldered.

NOTE: New designs should not use this form of transistor connection unless no other form of connection is possible.

B – Tag strips with turret lugs or flat tags

1. WIRE LINKS

Wire links between adjacent turret lugs must be made in 0.560 mm (24 SWG) BTC wire. When three or more lugs adjacent to one another are to be linked, the link wire is to be wrapped 180° round the first lug, 360° round the centre lug or lugs and 180° round the last lug of the group in a continuous wire.

With flat tags, BTC links between three or more tags shall have lay-on joints on the centre tag or tags but the first and last tag must be terminated as in Section 4B.

Link wires must be fitted first at the base of the lug, followed by the component wires, and then finally the external interconnecting wires at the top of the lug.

Insulated link wires passing under the board must not lay across the riveted end of other turret lugs.

2. COMPONENTS

Components may be fitted so that they lie in contact with the board but only if they can all be fitted in this manner on the board.

Components must be fitted so that their markings are readable and all in the one direction if at all possible. Basic rule for colour codes is, 'read from top to bottom or from left to right'.

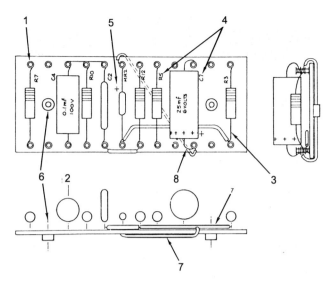

KEY
(1) Tag strip with flat tags must conform to the above pattern but connect components between inner tugs
(2) Components to be central between tags and if clear of board clear by between 3 and 6 mm
(3) Component wires not to be under tension
(4) No obstruction of component references on board
(5) Component polarity to be clearly visible
(6) Hank bushes must not be obstructed
(7) Insulated wire links may run over or under board but must originate from top
(8) Sleeve too short

Fig. 2.3 Typical component mounting board or tag strip.

Components over 14 g in weight must be secured directly to the tag strip by a mechanical method designed to remove the component weight from the connecting leads.

A spring clip, screwed clamp or PVC-covered nylon tie are all acceptable methods.

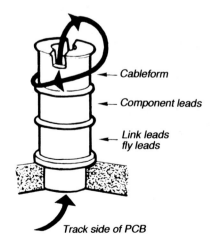

Cableform

Component leads

Link leads fly leads

Track side of PCB

Fig. 2.4

Ensure that the connections have been made to the specified part of the terminal.

The common procedure shown is designed for easy maintenance. Cableform leads are connected to the top groove to enable the subassembly to be easily removed. Components are connected to the centre groove to permit easy removal from the subassembly. Link wires and fly-leads are connected to the bottom groove.

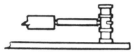

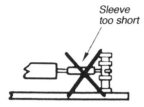
Sleeve too short

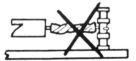

Fig. 2.5.

- Ensure that sleeves are fitted where specified. They are used where there is a danger of short-circuits
- Check that the sleeve is the correct length, colour and type
- Check the insulation trap between the sleeve and the terminal. This should be about the diameter of the sleeve
- Ensure that the sleeve is not damaged

C – Printed circuit boards

1. GENERAL

Unless otherwise stated by the design authority, components must be fitted so that when possible, downwards pressure on the components will not cause a force to be applied to the soldered joints at the PCB solder pads which would tend to make the copper track tear away from the surface of the PCB.

All components which exceed 14 g in weight must be securely mechanically fixed to the PCB and not rely on the termination wires to hold them in position.

The wire terminations must be preformed by careful bending (no sharp bends closer to the body of the component than 1.5 mm) so that the preformed wires easily go straight down through the mounting holes without undue pressure or distortion and the component lies midway between the supporting points. This statement applies equally to any non-electrical supporting tag or lug which may be provided for mounting purposes.

Under no circumstances must two components be connected directly in series unless the junction of their connecting leads is anchored firmly to the board by soldering to an intervening pin or pad.

Under no circumstances must two components be connected directly in parallel by soldering the leads of one component to the leads of the other. Both components must be mounted separately to the PCB.

To conform to the requirements in the first paragraph of the subsection transistors (or integrated circuits) must be fitted so that downwards pressure on the component body is not communicated as strain to the junction between the component body and leads or via the leads to the PCB track, the latter being very important with nonplated through holes.

Heat producing components (e.g. resistors over 1 w rating) must be mounted so that the heat producing body is clear of the surface of the PCB by at least 3 mm.

All components which are not directly in contact with the PCB must be mounted so that the bottom face of the component is parallel with the face of the PCB or the component stands vertically at 90° to the face of the board.

2. CIRCULAR-BODIED COMPONENTS

With the board viewed normally, i.e. top edge away from viewer, mount the component so that the colour code can be read left to right or top to bottom. Component value lettering and numbering should be visible on the top surface of the component wherever possible. Tubular bodied components mounted vertically should have their positive end at the top away from the board with the colour code reading towards the board.

3. FLAT-BODIED COMPONENTS

Where possible (subject to polarity requirements) capacitors must be mounted so that value markings are clearly visible and not obscured by surrounding components. Components should stand vertical and not be pushed over at an angle sideways. The case (encapsulation) of this type of component must not be cracked, cut, or in any way damaged especially where the leads emerge from the component body.

4. TRANSISTORS

Transistors in T05-style cases must always be mounted either with the component body pressed down in contact with a mounting pad which, itself is in contact with the PCB. When using the T05 case mounting pads must always be fitted under components which are fitted to PCBs with plated-through holes, or under transistors which are fitted with heat sinks. When ferrite beads need to be fitted directly to transistor lead-out wires, all the lead-out wires of the transistor so fitted must be sleeved with insulating material to:

- Prevent 'frettage' of the plating of the lead-out wire by the hard ferrite material (base metal of the lead is often ferrous and will rust if plating is damaged)
- Prevent electrical contact between lead and ferrite, some grades of which are electrically conductive
- Act as spacers to prevent track or lead strain

5. INTEGRATED CIRCUITS

Integrated circuits (in-line contact type) must be fitted down on the PCB so that the shoulders of the connecting tags are seated firmly against the board's top surface. The tags should then be cropped so that they project on the solder side by not less than 1 mm and no more than 1.5 mm.

6. INTEGRATED CIRCUITS (CONT.)

Integrated circuits in T05-style cases (8-, 10- and 12-lead type) should be fitted with a flat mounting pad held close to the component body while the leads are bent outwards, each by an equal amount, to fit the hole configuration of the PCB. The component, complete with mounting pad, must be fitted to the board so that the pad is parallel to the board face and clear of it by less than 3 mm or more than 6 mm. When more than one component of this type is fitted to a PCB they must all be fitted at an approximately equal height.

Important note

Certain integrated circuits are prone to damage due to electrostatic charge. These components are supplied from the manufacturer packed in special conductive tubes or with the connecting pins pushed into conductive black plastic foam; they must be handled with great care and not be removed from the special packing until *immediately* before the component is fitted (see Section H).

7. POTENTIOMETERS

Preset PCB mounted potentiometers with three pins projecting from a circular base, must always be mounted with their base (or lead shoulders) seated flat down on the board surface. The connecting pins will not normally be cut off but cropping is permissible provided that there remains at least 1.5 mm projection through the board. Cropping must take place before the joint is soldered and the cropped ends must be tinned.

8. SMALL TRANSFORMERS

Small iron dust cored transformers, not exceeding 14 g in weight, are to be fastened to the PCB surface by a small piece of double-sided adhesive plastic foam tape which has been cut to be the same size as the bottom face of the component. It must not project from under the component body nor be appreciably smaller than the flat base surface. There must be no strain on the transformer connecting leads which must be insulated to within 1.5 mm of the surface of the board. The leads must project through the soldered joint on the solder side of the board so that the wires may be checked for correct tinning.

9. WIRE LINKS

Where it is necessary to connect two points on a PCB with a wire link, it is to be made in 0.560 mm (24

SWG) BTC wire, providing it is not more than 19 mm in length and that there is no danger of a short-circuit. The link must run straight between the link points. For links longer than 19 mm and links which have a track on the PCB, the link must be made in 0.560 mm (24SWG) PVC insulating wire (pink). If the link is to be removable then the link points must be fitted with solder pins to which the wire must be soldered. All links should lie flat on the surface of the PCB.

10. FERRITE BEADS

Ferrite beads must always be fitted in such a way that they will be retained in their fitted position and will not move due to vibration. This is especially important when ferrite beads are fitted to transistor leads, large proportions of which are a very thin layer of gold plated on to a ferrous (Covar) core. When it is necessary to fit a ferrite bead to a transistor lead, all the leads of the transistor must be sleeved with insulated sleeves and the sleeving on the lead on which the bead is fitted must be a good enough fit in the bore of the bead to prevent movement of the bead on the lead. These precautions are aimed at preventing frettage of the lead plating by the hard ferrite with eventual exposure and rusting away of the ferrous core of the lead. Alternatively, the ferrite bead may be held in place with heat shrunk tubing.

11. PCB ASSEMBLY

The order of assembly of components to the PCB is generally in the following sequence:

(1) Preforming and hot tin dipping of component leads
(2) Assembly of components on to the board using a flap jig and using a pneumatic bend and crop tool to form and cut the leads
(3) Flow soldering of the assembled board
(4) Addition of further components after flow soldering.

12. ORDER OF COMPONENT MOUNTING

Components are mounted in the order shown on the assembly instructions. If for example, the larger components were mounted first, the small compo-

nents would fall out on reversing the fixture. Therefore, the smallest components are mounted first, then the slightly larger size and so on until the largest components are in position.

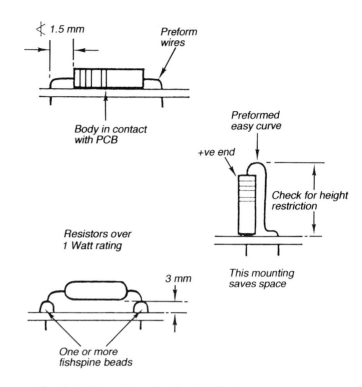

Fig. 2.6 Mounting circular bodied components.

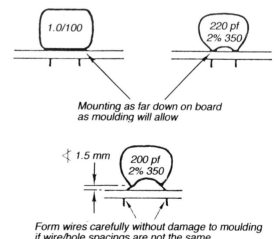

Fig. 2.7 Mounting flat bodied components.

NOTE: A neat application of a suitable potting compound such as RS® silicone elastomer or similar product around the base of the capacitor is an acceptable method of securing it to the PCB. However, it must be authorised by the design department or shown on the drawings.

Handling points

■ Do not flex a board as this could cause inter-mittent failures due to track breakage

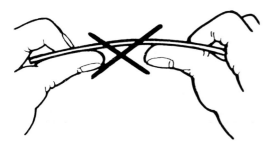

Fig. 2.8.

■ Do not handle a board with sweaty hands, as this could cause corrosion, or inhibit soldering. Use gloves, finger cots or other methods of handling circuit modules

Fig. 2.9.

When handling PCBs on which components have been mounted:

■ Do not stack modules on top of each other

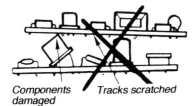

Fig. 2.10.   Components damaged   Tracks scratched

■ Do not apply pressure to components as this can damage the soldering

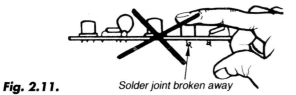

Fig. 2.11.   Solder joint broken away

The correct method of handling is to grip boards by the unplated edges.

Fig. 2.12.

■ Mount into tote boxes. PCBs with component must be transported using either company standard antistatic boxes designed to protect the modules or suitable antistatic foam or bubble wrap packing.

NOTE: Only lint-and sulphur-free paper, or if static precautions are required, suitable antistatic bags are to be used for packaging.

When assembling diodes, it is important that the polarity be observed, i.e. the cathode corresponds to the positive mark on the legend. The cathode of a diode can be indicated by a number of different methods. The most common of these are:

■ A coloured band
■ A plus sign
■ Type code colour coded at cathode
■ Arrow of symbol points to cathode
■ Cathode end has coloured bands or dot
■ Cathode end has type number printed commen-cing at cathode

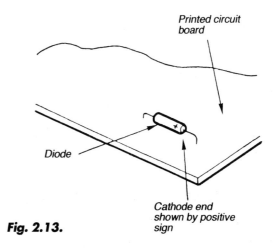

Printed circuit board

Diode

Cathode end shown by positive sign

Fig. 2.13.

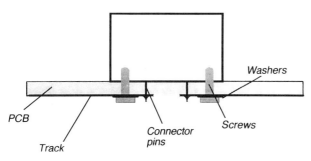

PCB

Track

Connector pins

Washers

Screws

Fig. 2.14.

Screws secure the body to the board.

NOTE: If the body of the component is required to be insulated, the correct insulating washers and inserts must be used.

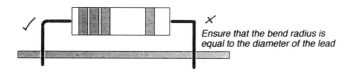

Ensure that the bend radius is equal to the diameter of the lead

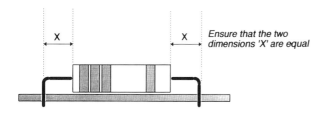

Ensure that the two dimensions 'X' are equal

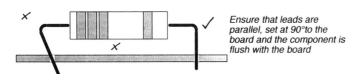

Ensure that leads are parallel, set at 90°to the board and the component is flush with the board

Fig. 2.15.

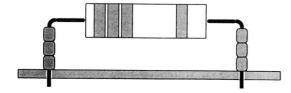

Where components require to be mounted away from the board to prevent scorching or allow additional airflow for cooling, they should be mounted on ceramic beads

Fig. 2.16.

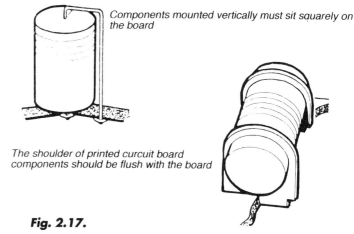

Components mounted vertically must sit squarely on the board

The shoulder of printed curcuit board components should be flush with the board

Fig. 2.17.

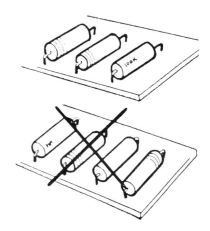

Fig. 2.18.

Forming of multiple resistors

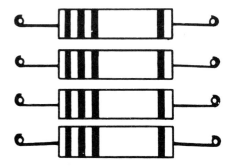

Fig. 2.19.

(1) Components to be level and equally disposed between pins
(2) All component leads to be formed on common side of pins
(3) Component leads to form 180° wrap round pins
(4) Rows of components form a neat line
(5) Components are in line with the component symbols on the PCB
(6) The identification of components reads from left to right in relation to the drawing

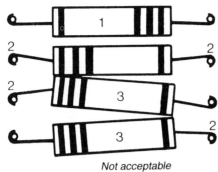

Fig. 2.20.   *Not acceptable*

(1)   Coding reversed
(2)   Leads incorrectly formed on pins
(3)   Components misaligned

Incorrect component bending centres

There should be a minimum of 1.5 mm straight wire after the component encapsulation ends before the bend in the wire. Less than 1.5 mm is a Class C defect. This applies to most components including resistors, capacitors, transistors and diodes. The radius of the bend in the wire should not be less than the diameter of the wire. Where the bend is made with a small radius the 1.5 mm is measured from the inside of the bend to the beginning of the wire body termination.

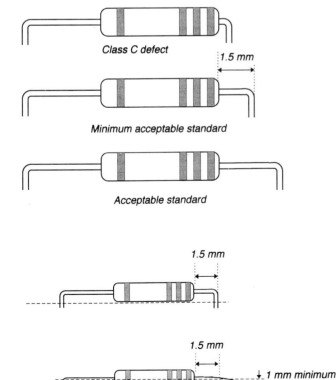

Class C defect

1.5 mm

Minimum acceptable standard

Acceptable standard

1.5 mm

1.5 mm

1 mm minimum

Fig. 2.21.

Where the bend is made with a much larger radius there should be 1.5 mm straight wire before the wire goes more than 1 mm from straight.

Incorrectly seated components

This component is a Class C defect because it is raised more than 2 mm. This is likely to raise it above the level of surrounding components and makes it liable to be damaged in normal handling.

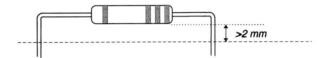

>2 mm

This is an acceptable way of mounting a resistor which for any reason has to be mounted clear of the board. This method and other similar methods do not make damage likely during normal handling.

This is the normal acceptable way of mounting small resistors. There are many other acceptable ways which can be used in which the component seats flat on the board.

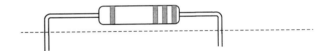

Fig. 2.22.

Damage to components

The illustration shows one of the many types of component which may be damaged. Damage to case that does not expose inner part of component is acceptable.

Fig. 2.23.

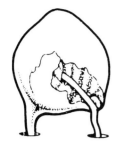

CONSIDERABLE DAMAGE
Damage likely to shorten life of
component
Class B defect

UNKNOWN DAMAGE
Effect of damage unknown but could
be considerable
Class B defect

Fig. 2.24.

Ensure that clips are used to support components where specified. It is recommended that components weighing over 14 g should be supported.

Ensure that the clip is secured to the board and that the component is held firmly in the clip.

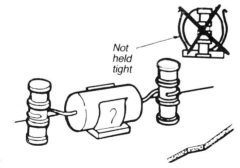

Not held tight

Fig. 2.25.

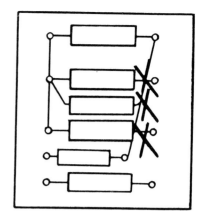

Fig. 2.26.

Check that terminal lugs are tight and that they have not been loosened by strain during the assembly, or by overheating during soldering.

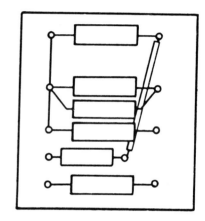

Fig. 2.27.

Forming tin copper wire to transistors for pins of 1.5 mm diameter or less.

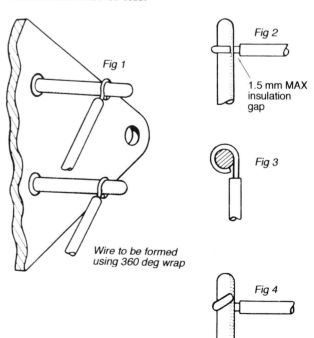

Fig 1

Fig 2

1.5 mm MAX
insulation
gap

Fig 3

Wire to be formed
using 360 deg wrap

Fig 4

Fig. 2.28.   *Not acceptable*

Damage to component wires

Component wires should be free from nicks, cracks and other damage which might cause a reduction in the cross-sectional area and subsequently the current carrying capacity of the wire. If, when viewed from any angle, a component wire is nicked or otherwise damaged so that its diameter is reduced by up to 20% this is a Class C defect. If the diameter of the wire is reduced by 20–25% this is a Class B defect. If the wire is reduced or damaged by more than 25% it is liable to failure and is unsuitable for use. This is a Class A defect.

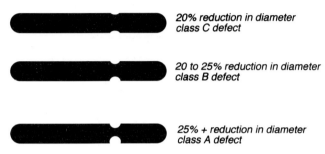

20% reduction in diameter
class C defect

20 to 25% reduction in diameter
class B defect

25% + reduction in diameter
class A defect

Fig. 2.29.

D – Chassis-mounted electrical components

1. ELECTRICAL BONDING

Unless otherwise specified by the design authority, all electrical components which are mounted directly on to a metal chassis, frame or panel and which themselves have a metal frame case or cover shall be securely electrically bonded to the chassis, frame of panel. Such bonding will usually be obtained by the bare metal to metal contact at the bolting face of the component and chassis, or equally via bare metal to metal contact through the securing screws.

If the bonding contact is made through a screw fixing to a chassis or panel having a painted finish then the chassis or panel must have the area of the screw seating concentric with the screw hole masked before the chassis of panel is painted. The masked area should be between 1.5–3 mm larger in diameter than the screw head. Spot facing away of the paint is only permissible if the panel or chassis is made of aluminium.

Where two items both having painted surfaces are required to be fastened together with the painted surfaces touching and are also required to be electrically bonded then a masked area concentric to each fixing hole and 1.5 mm larger than the diameter of the standard washer for the size of the fixing used shall be provided in each painted surface. On assembly of the two painted items a standard electrotinned brass washer shall be fitted between the painted surfaces on each fixing used, to take up the thickness of the paint and to ensure an adequate metal to metal contact at the masked areas.

If, for circuit design considerations, the above bonding methods are not considered sufficient by the design authority then the required alternative method must be explicit on the drawing and there must be no deviation from the instructions. Special earth bonding via scissors spring contacts is often found to be necessary at various points along wafer switch operating shafts in RF circuits and the position and efficiency of these contacts must be strictly observed.

2. ANTI-ROTATION DEVICES

Rotary switches and potentiometers designed for panel mounting which have a pin or lug provided to prevent rotation of the component body must have a hole or depression provided in the appropriate position in the rear face of the mounting panel to engage with the pin or lug so that this antirotation device can be effective.

3. RETAINING DEVICES

Electrical components (i.e. valves, crystals, plug-in relays, etc.) which are plugged into special holders mounted on the chassis must always be fitted with a positive retaining device which will prevent the component from jumping out of its holder under heavy shock. Acceptable devices are:

- Spring loaded circular can which is pressed down and turned through part of a revolution to lock over the component
- U-shaped spring clip which is pulled up and sprung over the top of the component
- Lugs or plate mounted on two spiral tension springs which may be pulled up and positioned over the top of the component to apply downwards pressure
- Retaining plate fitted to a pillar to form an inverted 'L' over the top of the component.

E – Power transistors and diodes

Some basic points to remember:

- All power transistors and diodes must be mechanically fixed to their respective assembly, unless shown explicitly on the drawings
- Correct contact between the component body and its associated heat sink must be maintained at all times
- All mechanical fixings should avoid applying stress to their terminating soldered joints
- All heatsinks used should be sufficient so as to maintain the transistor body temperature below the design level. NOTE: Full consideration must be taken of the equipment's environmental specification – temperature and humidity range, etc.
- Where a heat conducting compound or mica washers cannot be used, it is usually necessary to use a larger heatsink to guarantee correct heat transfer and, in some cases, to 'lift' the compound assembly clear of the PCB to allow natural air flow cooling
- Always ensure sufficient clear space around the component to enable correct cooling

Typical assemblies are shown below.

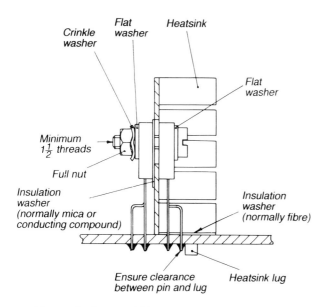

Alternatively, a self-lock nut can be used with a plain washer only.

Fig. 2.30 Transistor TO126 assembly.

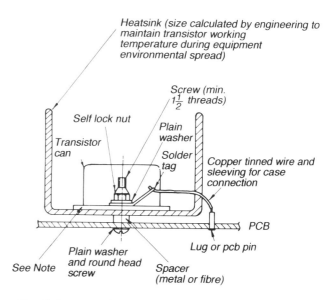

Fig. 2.31 Transistor TO3.

NOTE: The filling of either a mica washer or conducting compound is optional depending upon either the application or the customer's requirements. The emitter and base pins are soldered directly into the PCB *after* mechanical fixing.

1. DOUBLE HEATSINKS

It may be necessary to fit double heatsinks and the method of fixing can either be as above or, alternatively, as shown in Fig. 2.35.

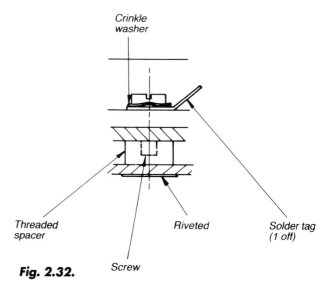

Fig. 2.32.

F – Flat pack assembly

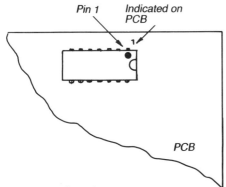

Fig. 2.33 Basic location.

The points regarding the fitting and soldering of flat pack assemblies are equally relevant to hand and machine soldering. The points mentioned here are not to be taken as the only methods to be used for special applications. The fixing methods should be checked with the supplier.

1. FLAT PACK ASSEMBLY

Main points to note

- Body of component lies flat to PCB
- Pins to be correctly located. NOTE: It is possible for a pin to be bent underneath the body and remain undetected – especially if automatic insertion methods are used
- All pins must be undamaged. NOTE: it is common practice to bend one pin slightly or, alternatively, to solder tack it in position to allow the PCB to be hand or machine soldered (see Fig. 2.37)
- Pin 1 on the component to coincide with pin 1 screen print identification on the PCB

Pins can protrude a minimum of 1 mm and a maximum of 3.2 mm.

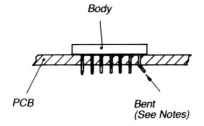

Fig. 2.34.

G – Surface mount assembly

In order to solder surface mount components correctly it is essential that the components are positioned accurately. The diagrams on p 46 show the correct placement of the four main types of surface mount components.

1. SOLDER STANDARDS

Introduction

It is worth repeating that, in view of the high reliability requirements of products supplied by the electronics industry, the establishment of a quality standard for soldered joints is necessary. The characteristics determined here represent the minimum requirements for the types of soldered joints that are typical for printed board assemblies, containing surface mount components.

As previously mentioned successful soldering depends, to a large degree, on a number of individual contributing factors, e.g. flux, waveform, reflow method, soldering irons, temperature, design of soldering leads, the shape and material of component leads, materials, finishes, etc.

Scope

These guidelines are designed to cater for the type of printed board assembly manufactured within the electronics industry. Other specifications (typically customer specifications) should be considered as they apply.

For the purposes of visual evaluation the examples of soldered joints shown and described apply equally to: nonplated through hole boards, plated through hole boards, and hybrids containing surface mounted components.

Criteria for the assessment of soldered joints (solderability)

Solderability be defined as: the ability of a surface or surfaces to be wetted with molten solder resulting in the formation of a smooth, continuous solder film or fillet.

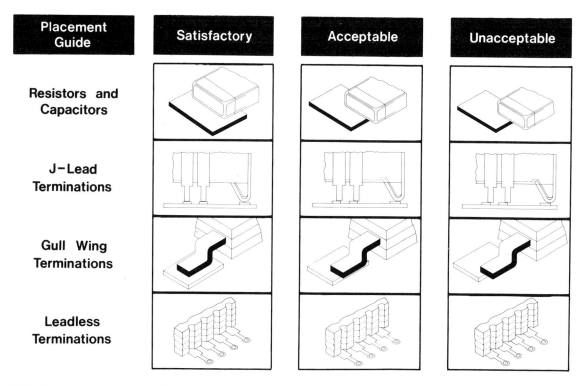

Fig. 2.35 Correct component alignment (reproduced by courtesy of Dimension 2 Technology).

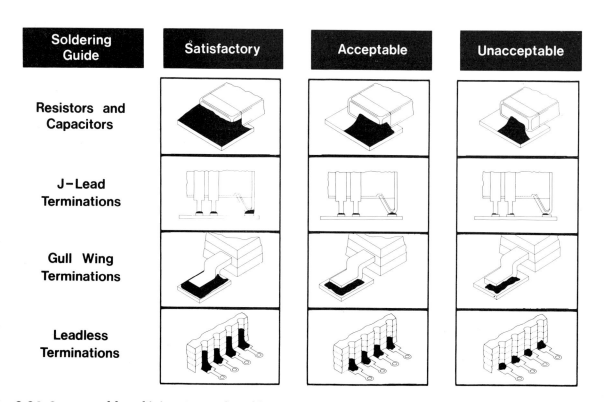

Fig. 2.36 Correct soldered joints (reproduced by courtesy of Dimension 2 Technology).

Poor solderability results when the surfaces to be joined are not in the correct combination. This results in nonwetting, or failure to alloy and gives rise to poor quality soldered joints.

Solder joints applicable to surface mounted components

(General)

It should be noted that with the surface mounted technology the solder joint may be the only means of attachment between the component and the board. Consideration must be given to this fact during design to ensure the ability to manufacture, test and inspect the board.

The visual standards shown in the illustrations are intended as an aid to determining the acceptability of the joints on the connected component. The diagrams provide an illustration of satisfactory, acceptable and unacceptable soldered joints and are marked accordingly.

The visual appearance of solder joints on surface mounted components may vary due to the type of process used to assemble and solder the components.

All boards should be cleaned prior to inspection to remove all traces of flux.

There should be no evidence of solder balls or slivers on the board. If present, the process must be investigated to determine the cause.

Components which are soldered in place using wave soldering will normally be glued to the board prior to soldering. Minimal adhesive on component terminal points is acceptable, provided that it does not reduce the minimal solder fillet. Evidence of terminal contamination by adhesive must be examined and corrective action taken before further production.

Any evidence of poor solderability or leaching must be investigated and tests for solderability on remaining stocks should be conducted.

Repair work on boards containing surface mounted components needs special consideration. Reference should be made to Section 7.

Resistor, capacitors with metalised terminations

Solder joints should exhibit a visible solder fillet between the pad and component termination. The fillet should rise to a minimum of 25% of the terminal, with evidence of good wetting. A continuous fillet should be visible around 75% of the metalised area and termination.

The outline of the termination should be visible in the solder fillet.

The maximum of 10% loss of termination (leaching) is acceptable. This is, however, an indication that a problem exists with the components or the process. Both should be examined for possible correction. If leaching has occurred, it will generally be visible on the corners or edges of the metalisation. Because of their construction capacitors may delaminate or crack and evidence of this will require removal and investigation of components and process.

No more than 25% of the component termination should overhang the pad, provided that the minimum clearance between conductors is maintained with components flat to the surface of the pad.

Self resistors and round leads

Solder joints on round contacts shall exhibit a visible fillet between the pad and component termination. The fillet should rise to a minimum of 25% of the terminal, with evidence of good wetting. The outline of the termination should be visible in the solder fillet.

A continuous solder fillet shall be visible around 75% of the termination.

SO and SOT packages

Leaded chip carrier solder joints on flat leads shall exhibit a visible fillet between the pad and component. The fillet should rise to the top of the lead with the outline of the component lead visible in the joint.

A continuous solder fillet shall be visible around 75% of the lead.

Minimal marks left by probes used to hold components in position during reflow are acceptable.

The heel fillet should be continuous between the heel of the lead and the pad, wetting should extend to a midpoint between the upper and lower bend as a maximum.

Any sign of nonwetting or dewetting requires examination of the boards and components for satisfactory solderability.

A solder fillet should be visible around 75% of the lead and should be visible to a minimum of half the lead thickness.

Leads may be raised off the pad surface provided this does not exceed two lead thickness, however, a good solder fillet must still be visible.

Any machine misplacement found, or damaged leads, must be investigated for correction in future production.

Leads may have a side overhang provided that this does not exceed 25% of the lead width and the minimum clearance between conductors is maintained.

Gull wing chip carriers should exhibit the same soldering requirements as other leaded devices.

Leads not designed to have wettable sides are not required to have side fillets. However, the joints should permit easy inspection of all wettable surfaces.

Inspection consideration should be given to the condition of all lead frames and their correct plane height. Any excessive variation should be investigated for causes of damage, whether it be assembly or poor packing.

Leadless chip carriers

Solder joints should exhibit a visible solder fillet between the pad and termination. A joint is not required where no pad is present.

The solder fillet should rise to a minimum of 25% of the metalised termination with evidence of wetting between castellation terminal and pad.

Overhang of the terminal is allowable to a maximum of 25% provided the minimal joint is present.

All termination points must have a tin/lead finish prior to processing. Components should be obtained in this condition from the supplier or be tinned inhouse.

Gull wing lead terminations

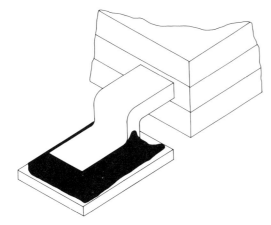

Fig. 2.37 Satisfactory. The solder joint should show satisfactory wetting between both the land and termination. Solder may cover the termination with the lead visible.

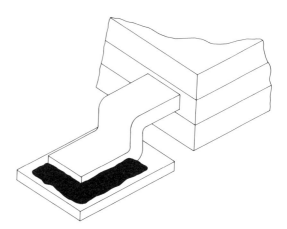

Fig. 2.38 Acceptable. A solder joint should be a minimum of 75% around the lead termination and a minimum of half the thickness (reproduced by courtesy of Dimension 2 Technology).

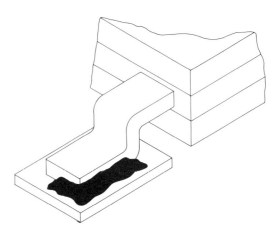

Fig. 2.39 Unacceptable. A solder joint with less than 75% around the component lead termination (reproduced by courtesy of Dimension 2 Technology).

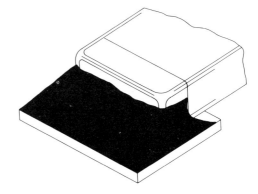

Fig. 2.40 Satisfactory. A solder joint should show satisfactory wetting between both the land and termination (reproduced by courtesy of Dimension 2 Technology).

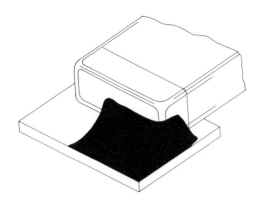

Fig. 2.41 Acceptable. The solder joint should be a minimum of 75% of the component and land termination (reproduced by courtesy of Dimension 2 Technology).

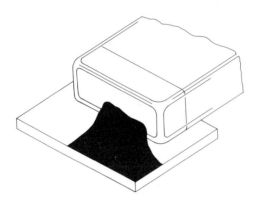

Fig. 2.42 Unacceptable. When a solder joint is less than 75% of the component and land termination (reproduced by courtesy of DImension 2 Technology).

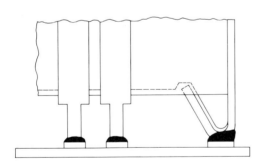

Fig. 2.43 Satisfactory. The solder joint should show satisfactory wetting between both the land and termination (reproduced by courtesy of Dimension 2 Technology).

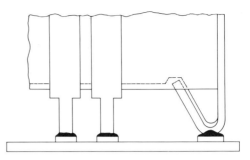

Fig. 2.44 Acceptable. The solder joint should be a minimum of 75% of the component and land, and a minimum of half the lead thickness (reproduced by courtesy of Dimension 2 Technology).

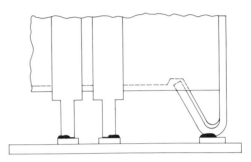

Fig. 2.45 Unacceptable. A solder joint with less than 75% of the component and land termination (reproduced by courtesy of Dimension 2 Technology).

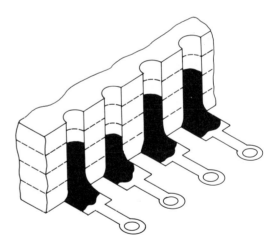

Fig. 2.46 Satisfactory. The solder joint should show satisfactory wetting between both the land and castellation termination (reproduced by courtesy of Dimension 2 Technology).

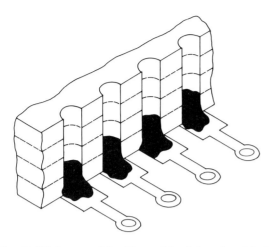

Fig. 2.47 Acceptable. The solder joint should show a minimum of 25% rise in the castellation (reproduced by courtesy of Dimension 2 Technology).

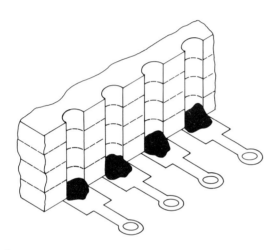

Fig. 2.48 Unacceptable. The solder joint shows no rise in the castellation (reproduced by courtesy of Dimension 2 Technology).

2. COMPONENT POSITIONING ON THEIR PADS

J lead termination

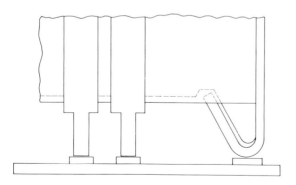

Fig. 2.49 Satisfactory (reproduced by courtesy of DImension 2 Technology).

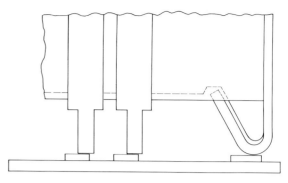

Fig. 2.50 Overhang on the lead termination is acceptable up to a maximum of 25% (reproduced by courtesy of Dimension 2 Technology).

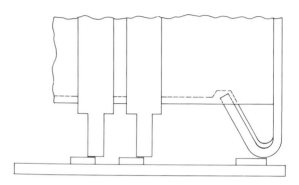

Fig. 2.51 Unacceptable. Overhang on the lead termination should not exceed 25% in any direction (reproduced by courtesy of Dimension 2 Technology).

3. LEADLESS CHIP CARRIERS

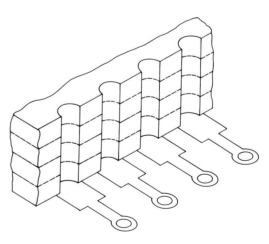

Fig. 2.52 Satisfactory (reproduced by courtesy of Dimension 2 Technology).

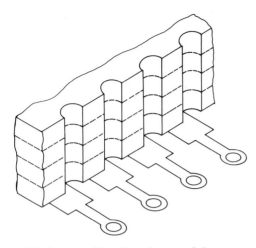

Fig. 2.53 Acceptable. Overhang of the castellation is acceptable up to a maximum of 25% (reproduced by courtesy of Dimension 2 Technology).

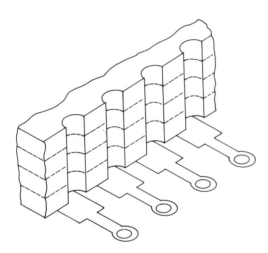

Fig. 2.54 Unacceptable. Overhang of the castellation shall not exceed 25% in any direction (reproduced by courtesy of Dimension 2 Technology).

4. RESISTOR AND CAPACITOR CHIP COMPONENTS

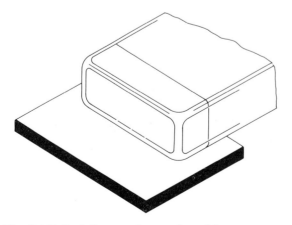

Fig. 2.55 Satisfactory (reproduced by courtesy of Dimension 2 Technology).

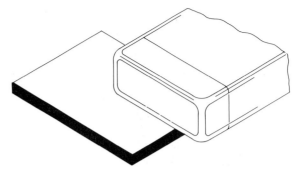

Fig. 2.56 Acceptable. Overhang of the termination is acceptable up to a maximum of 25% (reproduced by courtesy of Dimension 2 Technology).

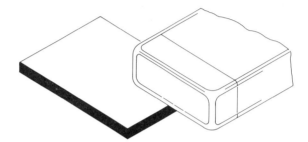

Fig. 2.57 Unacceptable. Overhang of the termination shall not exceed 25% in any direction (reproduced by courtesy of Dimension 2 Technology).

5. GULL WING TERMINATIONS

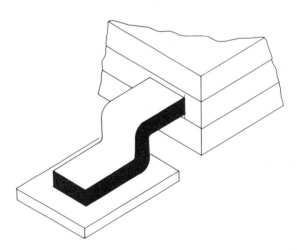

Fig. 2.58 Satisfactory (reproduced by courtesy of Dimension 2 Technology).

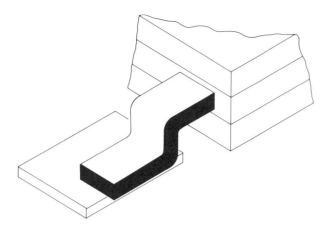

Fig. 2.59 Acceptable. Overhang of the termination is acceptable up to a maximum of 25% (reproduced by courtesy of Dimension 2 Technology).

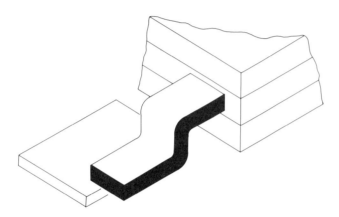

Fig. 2.60 Unacceptable. Overhang of the termination shall not exceed 25% in any direction (reproduced by courtesy of DImension 2 Technology).

6. QUALITY CONTROL POINTS

Assessment of the solder should be made with reference to each of the solder joints. Inspection of soldered joints should be conducted under conditions of adequate lighting with use of approximately × 5 magnification. If required, a maximum of × 10 magnification may be used in cases where the joint standard cannot be judged at × 5.

When examining solder joint wetting angles, consideration should be given to the board design and land areas as this may affect the shape of the joint. Depending on the process used for assembly this will also affect the appearance. Reflowed solder joints will appear to have less solder than flow-soldered joints. In cases where too much solder has been applied this may give the appearance of a bad wetting angle.

Where this is evident, the process of assembly should be examined. Excess solder, where it is widespread, may lead to reliability problems.

Classification of soldered joints

During the assessment of soldered joints by inspectors the following classifications should be considered.

(a) Satisfactory
This is a satisfactory condition which should be achieved and used as the standard for manufacture.
(b) Acceptable
This condition represents the maximum acceptable departure from the 'satisfactory' condition. Joints within this limit of deviation will not require reworking. Individual clarification accompanies each illustration.
(c) Unacceptable
This applies to an unacceptable joint condition which should not be reworked without the causes of the fault being established. Reworking may be possible after the assessment of the fault.
(d) Investigation of soldering process
The production department should examine the soldering process and the condition of the component termination points when the joint quality falls and remains within the standards of acceptability. It should not, however, be cause for rejection.
(e) Criteria for reworking joints
As referred to above, reworking is normally only considered on joints judged to be unacceptable. The reason why the joint is unacceptable must be established prior to commencing the rework since this may well have a bearing on whether successful reworking is achievable.

H – Static precautions

Precautions must be taken to safeguard components that are likely to be damaged by static electricity discharge *wherever* they are handled or likely to be handled. Throughout the manufacture of the components very costly precautions are taken to protect them from damage in every way including static discharge damage. All this effort is wasted, if manufacturing companies do not take precautions.

It is important to note that static discharge and

damage from static fields rarely causes catastrophic failures. Normally it will simply reduce the lifespan of a component to a few hours. It is not possible under normal circumstances to test for static damage. It is therefore necessary to take clear measures to ensure that components and equipment are not exposed to static discharge.

The following information is devised to highlight the areas of potential danger during a manufacturing sequence. NOTE: It is vitally important that components are protected by design engineers and technicians to avoid false information about the reliability and performance of equipment undergoing design trials.

1. SPECIAL HANDLING AREA (SHA)

The SHA consists of the following main elements:

■ Conductive bench mat
■ Conductive wrist strap
■ All linked to a star point
■ Conductive floor mat
■ Good earth (An earth completely separate from the factory mains earth is strongly recommended to ensure the complete elimination of earth loops or 'noisy' earths. The resistance to earth is normally provided by the equipment material itself. A typical combination is shown in Fig. 2.61.)

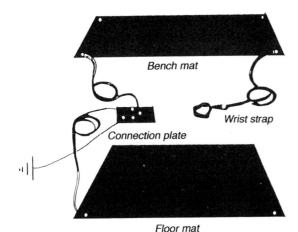

Fig. 2.61

Points to note in Fig. 2.61:

■ Standard bench mat $1.2\,m^2$
■ Standard floor mat
■ Wrist strap noncoiled 1.2 m (see Fig. 2.61)
■ Wrist strap coiled 0.5 m, 2 m extended length (see Fig. 6.62)
■ All the above are manufactured from materials

that either conduct or have a dissipative effect on static

■ When selecting bench mats it is important to use a proper dissipative mat. Use of conductive foil or metal sheet is incorrect. Dissipative mats will allow a statically charged object to discharge at a safe rate
■ If a charged object is put onto a conductor that has a direct connection to earth, serious damage may still be caused
■ Additionally the wrist strap must have a resistor in line of between $1\,M\Omega$ and $10\,M\Omega$ for the safety of the operator and the equipment
■ In areas where static precautions are in place personnel shall wear static dissipative coats
■ SHA should be identified by a station indication with the production line

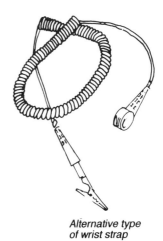

Fig. 2.62.   *Alternative type of wrist strap*

2. SOME TYPICAL ANTISTATIC PRODUCTION AIDS

■ Foam filled conductive component boxes, with lids, are normally used for integrated circuits
■ Storage bags should be
  ■ Transparent
  ■ Laminated/foam/metal
■ Static warning labels and tape
■ Subassembly transport carriers
■ Subassembly storage trays
■ Component storage bins
■ Component storage cabinet
■ Wrist strap
■ Conductive/dissipative mats and matting

The standard symbol for static sensitive components and precaution areas is shown in Fig. 2.63.

Fig. 2.63 Typical static sensitive indication.

Table 2.1 Electrostatic discharge (ESD) protective regime.

Manufacturing area	Precautions
Goods receipt Goods inwards Inspection	Components should be kept in manufacturers' packing. Sample inspection/test only. Any area where goods or samples are removed from static safe containers must be an SHA
Stores area	Booked in and stored as supplied. Where sample testing is performed, full ESD precautions should be used.
Kitting area	Components should be kept in supplied containers as far as possible. SHAs should be used if components need to be removed or containers changed
Subassembly	*Manual:* All operators full SHA precautions *Auto-insertion:* Machines earthed and operators use wrist straps or heel straps
Soldering	*Hand:* All operators full SHA precautions. NOTE: soldering irons should be earthed *Machine:* Machine should be earthed, operators should wear heel straps or wrist straps (bear in mind operator safety)
Equipment assembly	Full SHA precautions. NOTE: soldering irons should be earthed. Do not assume that components are safe because they are in a circuit board
Test	All test equipment earthed. Rework and components replacement to be carried out at SHA only
Inspection	All inspection stages work in SHAs
Service and repair area	SHAs to be used for component changes. Portable SHA to be carried by field engineers at all times
Quality assurance (QA)	Equipment to test safety and operation of straps mats and other equipment as required. QA to keep a log of equipment and when testing has been completed and when required

SECTION 3: ASSEMBLY OF MECHANICAL COMPONENTS

A – Introduction

With the variety of components and fixings involved, it is only possible to outline the general principles involved. However, the following important points should be noted by all personnel involved:

- The layout for the particular operation always details the fixings to be used. Care is required to use the specified screw lengths
- The electronics industry, and industry in general, is standardising on metric threads and 'Posidrive' heads wherever practicable. However, some older products still use imperial sizes and British Associated (BA) or other similar threads. Take care not to mix metric and imperial threaded items. Some of them are *almost* identical
- Always use the correct tools for mechanical assembly, in particular do not use pliers for tightening nuts
- Ensure fixings are tight but not overtight. Where necessary torque limiting tools should be used
- Taptite screws or screws with modified threads are used for some applications. Use care to distinguish between these types and conventional machine thread screws
- *Use care not to mark or scratch the surface of the material.* Dirty work benches are the biggest offender.

B – Screw fixing methods

1. GENERAL

Where two or more pieces of an assembly must be fastened together using screws and nuts the following standards must be observed:

- A screw/nut assembly must be tightened using hand tool torque only unless otherwise specified on the drawing. The nut should always be rotated on the screw whenever this is possible and a box or ring spanner should always be used in preference to an open ended spanner

- Appropriate tools of the correct size must always be used and the head of the screw, the nut and the surface of the parts being secured must not be damaged by the tools or the assembly operation
- At least two turns of the thread of the screw must project through the nut when a single nut is used

2. LOCKING

The screw/nut assembly must be locked by one of the following methods. The preferred methods being (1), (2) and (3) in that order:

(1) By the use of a 'crinkle' or 'wavy' washer fitted under the nut or screw head, whichever is turned during the tightening operation

(2) By the application of a proprietary locking compound specified by the designer. A typical product is Loctite Nutlock 242

(3) By the application of one drop of varnish or proprietary locking compound such as Loctite Penetrating 290 on the thread of the screw before the nut is assembled to it. If the screw is part of an electrical circuit the varnish or locking compound must only be applied after the nut has been fully tightened and then in a small quantity completely round the junction of the top face of the nut and the projecting screw thread. Varnish is confined to screws no larger than M5. If using a proprietary locking compound, refer to the manufacturer's instructions

(4) By the use of a second or 'lock nut' which is tightened against the first nut after this has been screwed down fully tight. The second or 'lock nut' should be a half nut unless otherwise stated on the drawing. The screw thread need not project through the lock nut by more than half a turn of screw thread, this being the minimum projection

(5) By the use of a 'Nylock' nut which must have a plain washer fitted under it. Nylock nuts must only be used when specified on the drawings

(6) If the above screw locking methods are not considered by the design authority to be appropriate for special cases, then other locking methods (e.g. tab-washers, prick punching, etc.) may be authorised and must be stated in the drawings

(7) Shakeproof washers (of the type designed to cut into the surface of the metal and screw head or

nut) must not be used. Spiral spring washers will not normally be used but may be authorised by the design authority if no other screw locking method is appropriate

(8) In the case of the 'bought out' subassemblies (e.g. relays, etc.), if shakeproof or spiral spring washers have been fitted by the manufacturer, then this locking method should be accepted and the existing washers should not be replaced unless they are supplied as 'loose fittings' (e.g. shakeproof washers supplied for mounting wafer-type switches

Screws, nuts and washers

To assemble component parts of equipment, screws, washers and nuts are used, the types commonly used are illustrated below.

UNF   Unified fine, follows the American thread form. Identification is by means of continuous rings on flats of hexagon

UNC   Unified coarse. Follows the American thread form. Identification is by means of contiguous rings on flats of hexagon

Metric   Universally used, different thread from British Standard. Standard adopted for all new designs in UK

BA   British Association. Different thread from the above

Types of heads and drives

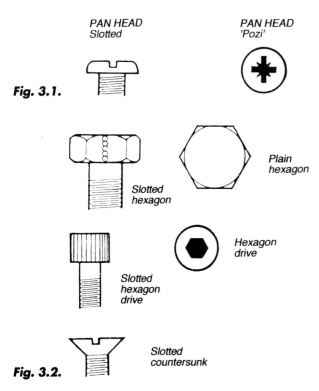

PAN HEAD Slotted

PAN HEAD 'Pozi'

Fig. 3.1.

Plain hexagon

Slotted hexagon

Hexagon drive

Slotted hexagon drive

Slotted countersunk

Fig. 3.2.

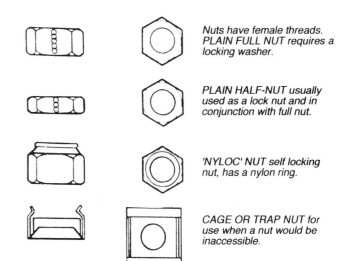

Nuts have female threads. PLAIN FULL NUT requires a locking washer.

PLAIN HALF-NUT usually used as a lock nut and in conjunction with full nut.

'NYLOC' NUT self locking nut, has a nylon ring.

CAGE OR TRAP NUT for use when a nut would be inaccessible.

Fig. 3.3.

Types of washers

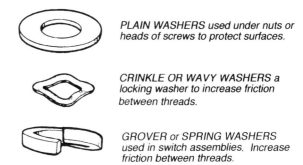

PLAIN WASHERS used under nuts or heads of screws to protect surfaces.

CRINKLE OR WAVY WASHERS a locking washer to increase friction between threads.

GROVER or SPRING WASHERS used in switch assemblies. Increase friction between threads.

Fig. 3.4.

Nuts

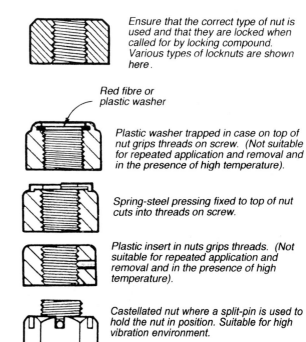

Ensure that the correct type of nut is used and that they are locked when called for by locking compound. Various types of locknuts are shown here.

Red fibre or plastic washer

Plastic washer trapped in case on top of nut grips threads on screw. (Not suitable for repeated application and removal and in the presence of high temperature).

Spring-steel pressing fixed to top of nut cuts into threads on screw.

Plastic insert in nuts grips threads. (Not suitable for repeated application and removal and in the presence of high temperature).

Castellated nut where a split-pin is used to hold the nut in position. Suitable for high vibration environment.

Fig. 3.5.

Electrical joints

Solder tags and flag tags may be secured to panels, etc.
Check that:

- Slots of screw heads are undamaged
- Nuts are tight
- Screw threads protrude past nuts, minimum one and a half threads

Also check whether there is electrical continuity to the panel.

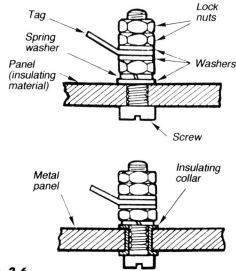

Fig. 3.6.

Heavy duty electrical connectors

Ensure that:

- The faces of the busbar and take-off connection are clean and flat, thus ensuring that maximum contact area is maintained
- Nuts and screws are fitted with the correct washers and are tightened to the specified torque

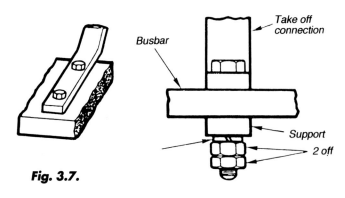

Fig. 3.7.

Pop rivets

Check the following on pop riveted assemblies:

- That the rivet is not split
- That the holes drilled in metal are not so large that they will cause misalignment
- Broken-off head has been removed from the assembly, and is not allowed to drop into any mechanism or electrical circuit

Check the riveting gun periodically to ensure efficient action.

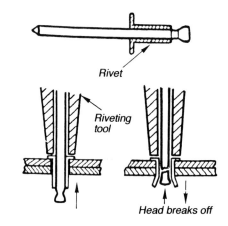

Fig. 3.8.

Rivets

Example of good riveting.
Check that:

- The parts riveted are in close contact
- The head of the rivet is of the correct form

Five examples of bad riveting

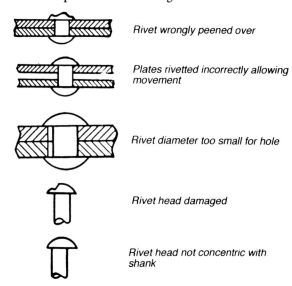

Fig. 3.9.

Screws

Method of measuring screw lengths.

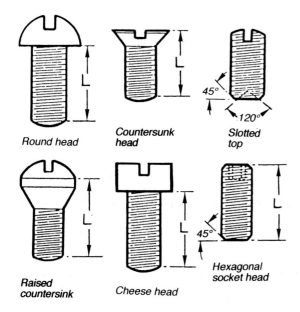

Fig. 3.10.

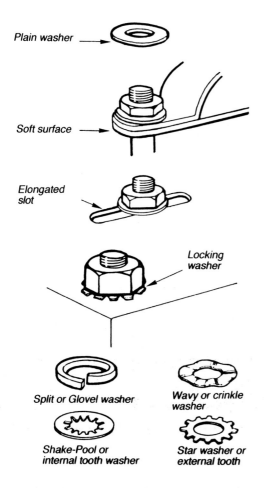

Fig. 3.11.

Fixing tag boards to chassis

Tag boards are mounted by means of brackets.

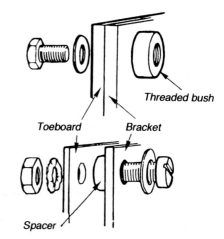

Fig. 3.12.

Fixing subchassis to chassis

(1) Mount subchassis by means of bolts or screws
(2) Threaded bushes, commonly called *Rosan nuts* attached to the main chassis are used where it is not possible to use nuts

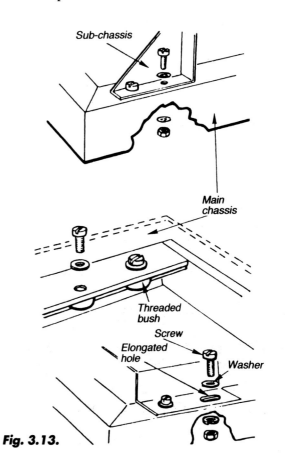

Fig. 3.13.

When fitting a screw into a tapped or 'blind' hole a crinkle washer is used under the head of the screw. If a 'P' clip is used then a plain washer will be in contact with the plastic.

Screws passing through elongated holes have plain washers fitted, either under the head of the screw or under the nut, this is to provide a larger load bearing surface than the screw or nut can supply.

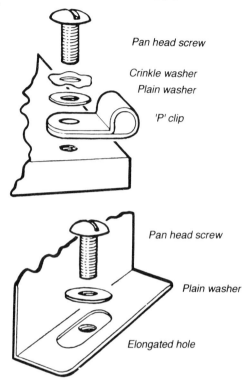

Fig. 3.14.

A correctly assembled screw with its associated washer(s) and nut will have a maximum of three and a minimum of one-and-a-half threads protruding from the nut, this is to ensure that all the load bearing threads are effectively engaged, with no undue stress imposed on the threads of either the screw or nut. 'Nyloc' nuts have a minimum of one thread protruding.

Attention must be given to the condition of the head of the screw or bolt head and its nut. They must show no evidence of burrs either to the screwdriver slots or to the flats of a hexagon, similarly the plating must be undamaged. All parts of the fixing assembly must be in close contact i.e. not held off its seat by burrs or foreign material. When a solder tag is used as part of a fixing assembly and the structure is of plastic, fibre glass reinforced plastic (GRP) or any insulating material, then the solder tag will have a plain washer placed between itself and the insulating material. This is to prevent the formation of a compression joint which can arise due to thermal expansion or shrinkage of the insulating material.

Grub screws

These are screws without heads in the normal sense. The drive is in the minor diameter, used for securing knobs to spindles.

Two types are used: cup, and point. The cup is tightened first, on the flat of a spindle then locked with a point grubscrew.

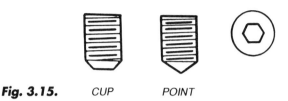

Fig. 3.15.   CUP   POINT

Helicoil screw thread inserts

These are female rolled threads formed in coils of stainless steel, diamond shaped wire. They are used to line a tapped hole. In a properly prepared hole they form standard threads of the original diameter. They are mainly used in aluminium castings and plastic to prevent wear and damage.

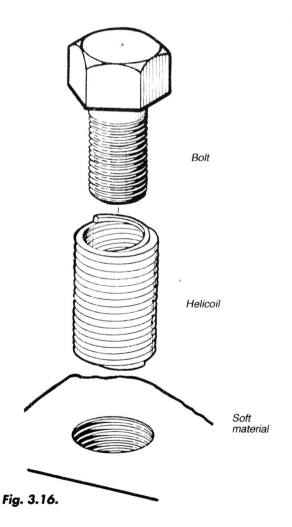

Fig. 3.16.

Locking washers may be used against the nut, or the screw head.

Plain washers are used to protect panels from damage caused by spring or shakeproof washers.

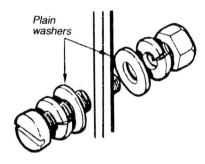

The use of locking washers is usually mandatory where the screw is screwed into a tapped hole.

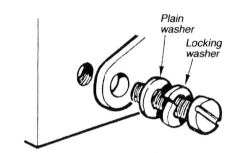

Fig. 3.17.

Ensure that all nuts and screws are tight. Some nuts and screws are tightened with torque spanners. It is important that these are correct, neither over-nor undertightened.

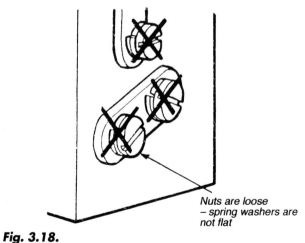

Fig. 3.18.

Locking varnish

Ensure that the specified locking varnish is used as required. The varnish should usually be applied to the screw thread before assembly.

Fig. 3.19.

Check for broken seals. A break in the varnish around the screw head indicates that the screw has been moved since the varnish set, and is no longer being locked by it.

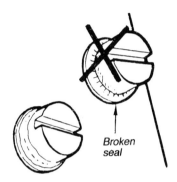

Fig. 3.20.

3. SOFT SURFACES

If the screw head or nut of a screw/nut assembly bears directly on to a soft surface (e.g. painted panel, insulated board, plastic component, etc.) then the soft surfaces must always be protected from damage by the direct pressure of the screw head or nut by the inclusion of a plain washer under the head of the screw or nut in direct contact with the soft surface, irrespective of the locking method used.

4. SLOTTED HOLES

A plain washer must always be fitted directly in contact with the item being secured when the hole in the item is slotted, irrespective of the locking method used.

5. HANK BUSHES OR CAPTIVE NUTS

NOTE: The screw must enter the threaded part of the hank bush by a distance not less than one-and-a-half times the full diameter of the screw. The screw must not normally project through a long hank bush when screwed completely home.

6. BLIND THREADED HOLES IN COMPONENTS

NOTE: Initial design should avoid this method of assembly if possible but if unavoidable 'Helicoil' inserts must be used in blind holes required to be tapped into aluminium.

The screw must enter the threaded part of the blind hole by a distance not less than one-and-a-half times the diameter of the screw and there must be at least 5 mm of usable thread clear of the end of the screw at the blind end of the tapped hole when the screw is fully tightened.

7. THROUGH THREADED HOLES IN COMPONENTS

NOTES:
(1) Initial design should avoid this method of assembly if possible and it must not be used in aluminium or insulated board. If unavoidable with other materials the material must be at least as thick as a full nut for the sized of the screw to be used.

(2) Short spacers or support pillars (up to 38 mm long for use with M3 screws) should always be specified with untapped clearance holes where this is possible and should be made from round section bar material. However, if due to design or assembly considerations, a spacer or support pillar is required to have a through threaded hole, then the spacer or support pillar must be made from hexagon (or square section) material never from a round section bar.

8. COUNTERSUNK HEAD SCREWS

NOTE: Except for references to 'fitting washers under the heads of the screw', subsections 1–7 apply equally to countersunk head screws. Countersunk head screws must only be used where a correctly countersunk hole is provided to take the screw head.

A countersunk screw head must not project above the surface of the material but it may be up to 0.25 mm below the surface if the screw head max-

imum diameter is not greater than 25 mm or 0.5 mm if the screw is over 25 mm diameter.

A countersunk screw should not be used if the thickness of the material in which the head of the screw is seated is less than 1.25 times the maximum depth of the screw head.

9. ROUND HEAD, HEXAGON HEAD, RECESSED HEAD SCREWS, ETC.

These screws will only be used where the form of their head is essential for correct operation of the equipment and cheese head or countersunk head screws would not be satisfactory.

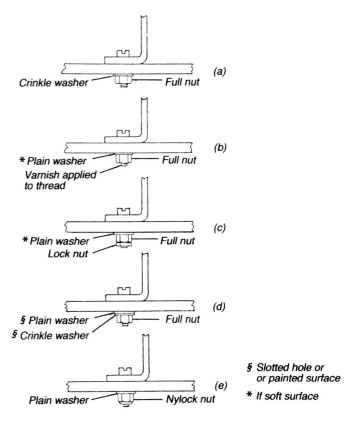

NOTES
(1) Assemblies must be tightened by rotation of nuts where possible
(2) Position of screw and nut may be reversed if necessary to facilitate (1)
(3) If it is not possible to tighten the assembly by rotation of the nut then a plain washer must be fitted beneath the screw head for protective purposes
(4) Screw projects through nut by minimum of two threads except at (c) which is minimum half thread

Fig. 3.21 Screw fixing and locking.

C – Riveted and swaged assemblies

1. RIVETS

Two or more components of an assembly which are fastened together by means of a rivet or rivets or swaging are to be considered to be a permanently fastened assembly and this form of fastening must not be used unless it is considered that the items being joined will never have to be parted. The following factors apply:

- Riveting is limited to small size rivets not exceeding 3 mm diameter and they are formed cold by hand or by single stroke pneumatic riveters
- There must be no burrs on the items to be riveted, their mating surfaces must fit flush together and be held firmly in this position during the operation
- The normal hole size for a rivet, should be the rivet diameter plus 0.03 mm (min.) and 0.7 mm (max.)
- No chemical processing must take place after an assembly has been riveted. Materials required to have plated or chemical dip finish must only be riveted after plating or dipping
- All rivets must be plated or tinned before use, excepting aluminium
- Damage or bruising of the plating caused by the riveting operation must be protected by the application of varnish to the bruised areas
- The head of a countersunk rivet must fit the countersunk hole with the surface of the head flush with the surrounding metal and with no appreciable holes or gaps round the periphery. The formed head of the rivet, if countersunk, should be concentric with the rivet shank, reasonably fill the countersink, and not project above the surrounding metal. If not countersunk but forming a projection, it must be neatly rounded and concentric with the rivet shank and sitting down all round to form a reasonable 'D' profile
- Hollow-type rivets must hold the materials firmly in contact with no signs of swelling of the shank forcing the materials apart. (This can occur if the hole is too large for the rivet.)
- The circular axis of the rivet must be at right angles to the face of the materials and both ends of the rivet must have swaged down squarely all round.

2. SWAGING – HANK BUSHES, INTERNALLY THREADED INSERTS, SPACERS, ETC.

This type of fitting must always stand at right angles to the surface of the material into which it is swaged and be anchored by the swaging firmly enough to withstand a reasonable amount of sideways force or any rotational force encountered when fitting or withdrawing the screw with which it mates.

The swaging must be even all round, fitting the countersink provided and not projecting more than 0.13 mm above the surrounding surface. There must be no splitting or cracking of the swaged metal and the threaded bore must not be obstructed or damaged. A screw entered into the threaded hole must stand at right angles.

Items must be plated before swaging and any bruising of the plating must be protected by a careful application of varnish. Care must always be taken that the application of protective varnish to hank bushes, etc. which are used to secure covers and screens does not adversely affect electrical or RF screening. Countersunk head screws should not be locked with varnish into hank bushes or swaged fittings if the screw is used to secure removable items such as access covers.

SECTION 4:
SOLDERING

A - Soldering

1. NATURE OF A SOLDERED JOINT

Solders are metallic alloys of lower melting point than the metals that they join. They act by:

(1) Flowing between the metal surfaces which remain unmelted
(2) Completely filling the space between the surfaces
(3) Alloying to the metals they are flowed to
(4) Solidifying

Condition (3) distinguishes solders from lead and other metals which can effect joining of roughened or undercut surfaces by penetrating and solidifying.

Solders do this, but in addition they alloy to the metal surface with a strength comparable to that with which a piece of solid metal holds together, that is, by chemical forces. Solder being attached by such forces, cannot be mechanically prised off the metal surface leaving it are in its original state, as is normally possible with a metal such as lead which has been cast upon a surface.

Examination reveals that in the action of wetting metals, such as copper and brass, the metal and the tin in the solder react together forming a layer of a chemically distinct intermetallic phase called an alloy.

Further indication of the chemical nature of the solder bond is afforded by the continuing growth of this compound layer so long as the joints are maintained at the soldering temperature. Solders are unable to wet a metallic surface which, having been exposed to air, has already entered into chemical combination with oxygen forming an oxide film.

The essential feature of a soldered joint is that each of the joined surfaces is wetted by a film of solder and that the two films of solder are continuous with the solder filling the space between them.

Extensive investigation by leading research institutions has led to the following conclusions:

- Coating of tin or alloys of tin with lead, zinc, cadmium or silver gives excellent initial soldering characteristics

- Hot-dipped tin parts are pre-eminent for soldering quality

- If easy soldering is the only criterion, tin alloy coatings have no advantage over pure tin

- With a brass base, deterioration may occur in the soldering properties of coatings after storage due to zinc diffusion and subsequent atmospheric moisture. A 0.0025 mm copper flash beneath electroplated tin or alloy coating or a 0.0015 mm thick nickel layer under hot-dipped tin alleviates this difficulty. It is also possible to solder to nickel plate

- A copper base reduces soldering properties owing to a loss of coating as tin/copper forms an intermetallic compound.

- A 0.0025 mm thick nickel undercoat gives greater ease of soldering and no deterioration in storage

- A copper flash of 0.0025 mm generally enhances initial solderability

- Where the coating system has to be chosen for reasons other than simple attainment of good soldering and the possibility of deterioration has to be faced, storage should be an ambient or preferably at 25°C to prevent condensation and it should be for the minimum possible period, this applies particularly to brass and copper work. Humidity and atmospheric contamination should be kept low, perhaps by storing the coated work in inert containers containing a desiccant

- Unduly high temperatures form excessive intermetallic compounds and lower joint strength, the contraction upon solidification is severe in solders with low tin content and is sometimes sufficient to cause cracks; this defect is termed hot-tearing

- Attempts to solder thin metal plate such as silver, gold or alloys of silver or gold may cause the plate to migrate into the solder or remove the plating completely. This can cause poor joints and possible corrosion over time

- Soldering over gold gives rise to the risk of brittle joints

2. IMPORTANCE OF A SOLDERED JOINT

In view of the many hundreds of hand-soldered joints that occur in most assemblies, each joint must be considered as a most important component. These important components are manufactured on the spot by operators when they make soldered joints. Unless suitable solders and solderable materials are provided, the operator and the joint itself can be at a disadvantage.

A considerable amount of difficulty has arisen

because various cable manufacturers have introduced many types of wire, some having tin plating to different standards and thickness, some are silver plated also at different thickness, and some nickel plated, together with various types of insulating materials which when heated can give off vapours which affect the joint formation. The problem is increased by component manufacturers providing lead-out termination with basic materials of copper/brass, nickel/iron, bronze and as many different types of plating.

3. SOLDERING AND THE IRON

Hand soldering has for many years taken a back seat to the development of more exotic metal joining techniques: resistance and parallel gap soldering, welding, infrared and wave solder machines comprise some of the more notable of these.

The advent of the space age with its expensive, complex, lightweight, high component density packaging has created increasingly stringent demands on this common hand tool. These demands are presented by the need to minimise the risk of failure – for even human life as well as time and costs are in the balance.

Sensitive electronic assemblies require that the thermal characteristics of an iron be carefully balanced and properly matched or fitted to the job. This matching should be carried to whatever economic level the product warrants. Soldering in the aviation or military environment is far more critical than the commercial variety. In the latter, wider temperature ranges or levels can be used successfully.

4. PRODUCTION SOLDERING

Production soldering is a continuous type of operation where skilled personnel perform the same or similar soldering operations, respectively, during the day. Production soldering operators have been known to produce as many as 30–60 connections per minute, or 12 000–25 000 connections per day.

5. INTERMITTENT OR JOB SHOP

Intermittent or job shop comprises the type of work where the iron may be idle for prolonged periods of time while the operator performs some wiring tasks.

In this case the iron should not overheat and must be readily available for soldering without tip tinning or dewetting problems. The operator usually solders three to ten consecutive connections, and returns the iron to the holder. Most repair positions would fall into this category.

6. DWELL AND TEMPERATURE

Many people insist on elevated temperatures to ensure fast production soldering rates, but in the process lose sight of the additional costs incurred in rework of connection, replacement of damaged heat-sensitive parts, removal of solder and flux splatter. Elevated temperatures also have adverse effects on tip and element life, and may cause tip dewetting problems. The proper iron supplies adequate heat to perform the solder task, repeatedly, and with complete utilisation of the flux. It produces shiny connections with good dihedral angle and tensile strength and a minimum of repair work.

Every solder connection is a by-product of two factors: time (dwell) and temperature. If temperature is controlled to a narrow thermal zone, meaning that the iron is matched to the task, then the only variable is the dwell. This makes training much easier and ensures better quality in the soldering operations.

7. THE IRON

The first considerations when selecting a soldering iron are the mechanical attributes. It is important that the iron has proper weight, length, balance and directional control (proximity of handle to tip), handle colour, etc. Substantial data has been compiled in various ergonomic studies (human engineering) and biomechanics publications verifying that these factors have a decided effect on fatigue and safety.

8. WATTS AND MASS

Wattage is one of the greatest misnomers in the electronic industry. For years soldering irons have been selected on the basis of wattage alone. Comparison of irons on a wattage basis could possibly lead to false conclusions. Instead, the really important factors to be considered (and those having a great effect on temperature) are:

- Mass of tip – large mass has high reserve
- Length of tip – short tips are more efficient
- Frontal mass of iron – the greater the mass, the lower the idling temperature of the iron; also the iron will be more stable under load because of greater heat reserve capabilities

Ideally, it is best to use the smallest, shortest and lightest iron with the thermal capacity to do the soldering task. Excessive weight promotes premature fatigue. Excessive length will cause the iron to be unbalanced, with the handle being too far from the tip for adequate directional control.

Colour of the handle has taken on a new importance, since studies have proved that certain colours are more visible than others at the edge of a person's peripheral vision. Use of the proper handle colours promotes safety and efficiency, there is less chance of missing the iron pick-up. The handle should have an adequate flare to prevent an operator's hand from sliding onto the hot case.

9. TYPE OF TIP

The choice of tip is as important as the selection of the proper iron. The following are some basic considerations:

- Select the tip shape which will give maximum contact area between tip and work. This will promote greatest transfer of heat
- Select the shortest reach to ensure fastest heat transfer from the element
- Use the shortest taper possible for better heat retention properties; long tapers reduce the cross-sectional area limiting good thermal conduction
- Use the largest diameter possible, turned down tips result in higher tip temperatures and lower heat reserve, reducing work output
- Tip life can be increased by proper design. A general rule to follow is to select a tip having the shortest length, largest diameter and largest point face that is consistent with accessibility of work.

For a hard-to-solder joint, design a point shape to closely match the configuration of the connection, complete with groove, notch or clearance hole to increase contact area. This type of tip should be specified in a long-life nonfreezing design. Tips may be selectively immunised to limit the tinned area. This will prevent carrying of excess solder to the work or to prevent bridging of solder between closely spaced terminals. Tips are frequently tinned on one side only and immunised against further spread of solder.

10. CONTROL OF TEMPERATURE AND VOLTAGE

Temperature measurements are made using an Adcola iron/constantin thermocouple loop, or other calibrated thermocouple. This type of loop has been found to provide an accurate and reliable readout of temperature. Variance in line voltage will create changes in tip temperature and may be a reason for a high soldering reject rate.

To ensure high reliability of the soldered connections all variables must be eliminated, the proper equipment specified, operators must be trained and continual supervision should be maintained.

11. THE FUNCTION OF THE FLUX

Most tin/lead solder alloys start melting at 183°C and liquid solder with a clean surface will wet clean metals but will not wet metal oxides. Unfortunately, practically all metals, no matter how thoroughly they are cleaned, oxidise immediately they are exposed to the atmosphere. For proper wetting, therefore, a chemical cleaning agent (or 'flux') must be applied which will remove oxide immediately in front of the advancing solder. Consideration of how this is to be achieved leads to the following requirements:

- The flux must be liquid at the soldering temperature. At room temperature it may be liquid, solid or pasty as convenient
- It must be capable of wetting both metal and metal oxides
- It must provide a reducing cover or at least prevent the access of atmospheric oxygen to the newly cleaned surface
- The flux residue after soldering must be readily removable or noncorrosive
- The flux must be capable of rapidly dissolving the metal oxide without detrimental attack on the metal.

It must be emphasised that a flux which fulfils all these requirements will not necessarily be able to deal with a metal surface that is heavily contaminated with oxide or with grease.

The flux also has the important subsidiary function of acting as a heat transfer medium between the source of heat (generally a soldering iron), the joint member and the solder.

Though the electropositive metals, such as gold and platinum, are readily soldered and the strongly electronegative metals, such as aluminium, are not, the electrochemical series cannot be taken as an infallible guide to ease of soldering as the melting point of the work-piece and the nature of the surface contamination are also involved.

Since fluxes do chemical work on metal surfaces, it is reasonable to regard the work available from a given flux (the flux activity) as the product of 'potential' and 'capacity' factors, just as one considers the work available from a dammed-up river to be measured by the height of the dam (potential' times the quantity of water (capacity). For practical purposes these two factors of flux activity may be regarded as measured by the extent of solder spread (capacity) and the speed of solder spread (potential).

The capacity of a flux is dependent to a large extent on the quantity of halide (or acid in the case of metal working fluxes) that it contains; generally a high capacity, (aggressive), flux is used only on heavily contaminated work-pieces or when soldering with a flame, which is likely to waste a large proportion of the flux by burning it before it has acted.

In electronic assembly high capacity is undesirable and normally resin-based fluxes such as, Online® 60/40 tin/lead, or one of the Multicore® 60/40 range, when soldering small, heat-sensitive components, fast action and limited spread is required. The flux not only removes surface oxides, but also prevents them reforming during soldering. Joints made with this flux do not corrode – even after prolonged exposure to humidity. The flux residue is pure resin, impervious to moisture, hard and nonsticky. It avoids accumulation of dirt on soldered joints. The maximum halide content of the flux is 0.5% and the approximate flux content is 3.0%.

B – Preparation for soldering

1. DESIGN FOR SOLDERING

Parts should be designed with the requirements of the soldering process in mind so that it is easy to make good joints; instances are common where it is plain that no consideration has been given at the design stage to the question of how the soldering is to be done or whether it can be done efficiently.

A prime consideration must be to arrange that the joints are conveniently positioned for soldering. Joints should not be too close to insulation that might be damaged by heat. In this connection it is worth remembering that when a surface is pre-tinned the joint can be made more quickly at a lower temperature. Tin/lead, and their alloys the solders, are weak metals in comparison with the copper and brass upon which they are so much used. The aim of the designer should therefore be to avoid, as far as it is practical, dependence on the strength of solders and to avoid putting any stress on the joint.

Thermal expansion should be allowed for, particularly when metals are dissimilar. It is not possible to accurately assess the expansion during ordinary bit soldering operations, but a fair estimate can be made for dip soldering where temperatures are more steady, on the assumption that during soldering the parts reach 265°C above a room temperature of 20°C.

2. STORAGE BEFORE SOLDERING

It is desirable that articles should not be put into store after cleaning, but should be soldered immediately. When this is not practical, consideration should be given to conditions of storage. The storeroom should be clean and not open to dust from surrounding works, and the temperature and humidity should be maintained at a constant level, otherwise there may be heavy condensation from time to time.

Articles in store should be protected as far as is economically necessary and practical, because the cleaning of even slightly rusted, tarnished or corroded articles before soldering is difficult and costly. Protection is provided by keeping the articles in boxes, wrapped in sulphur-free paper. Articles may also be coated in another metal such as tin, solder copper, gold or cadmium either by electrodeposition or, when applicable, by dipping in molten metal.

C – General

1. HAND SOLDERING

Unless specified otherwise by the design authority the solder used for all joints made with a hand soldering iron shall be *flux cored solder wire* to BS 441 Type 1 (activated), or 60/40 tin/lead alloy to BS 219 Grade K.

2. MACHINE SOLDERING

The solder and flux to be used for machine soldering shall be: compatible with (BS 219) Grade A but with lower antimony and impurity limits.

3. EXTERNALLY APPLIED CORROSIVE FLUXES

Externally applied corrosive fluxes (i.e. Bakers fluid, 'Fluxite', etc.) must not be used unless clearly specified by the design authority. In this case the jointed item must be capable of being completely chemically neutralised and thoroughly washed on completion of the soldering operation.

4. EXTERNALLY APPLIED FLUX – ALTERNATIVES

RS® liquid flux and pure resin, either as a solid or dissolved in spirit are acceptable as externally applied fluxes for hand soldering or dipping operations.

5. PRETINNING

Components which form part of a soldered joint must be adequately pretinned and cleaned before assembly of the joint, and soldering of the joint assembly must take place as soon as possible after tinning and assembly.

Joint components which have been gold-plated must be pretinned before assembly. Clean silver plating does not require pretinning.

6. PRECAUTIONS

The soldering operation must not result in any form of heat damage to surfaces or components in the immediate vicinity of the soldered joint.

Details of special precautions to be taken to prevent heat damage during soft soldering operations must always be specified in the operator's work instructions.

7. MECHANICAL STABILITY

A soldered joint must not normally rely entirely on the solder to keep the jointed items in contact. All joints must be designed and arranged to have a fair degree of mechanical stability before the application of the solder alloy.

8. 'LAY-ON' JOINTS

In certain circumstances 'lay-on' joints may be specified by the design authority if there is no alternative, but such joints must always be free from mechanical strain.

9. BURNT FLUX AND DIRT

The finished joint must be free from dirt and burnt flux residue. Minor traces of flux on the joint are admissible but heavy concentration must be removed and all traces of flux splatter and splashes of solder must be cleaned from the surrounding surfaces. See Section 5 for cleaning methods.

10. TRIMMING OF WIRES

The 'tail end' of wires projecting from soldered joints must not be cut off below the solder joint, after the joint has been soldered. Such action can strain the joint.

11. APPEARANCE OF FINISHED JOINT

The solder in a joint must present a bright tin finish smoothly and evenly coated over the entire joint area.

A dull matt or crystalline appearance to the solder surface usually indicates that the components of the joint have moved in relation to each other after the solder has cooled below the liquid point and the solder film is no longer solid. This is not acceptable.

The base metal of the joint components (i.e. bare copper, etc.) must not be visible in any part of the joint.

The outer edges of the jointing solder must blend in with the tinning on the surface of the components in a smooth concave curve and not cease abruptly to form a ragged outline or a convex 'blob'. The junction between the jointing solder and the component tinning should be hardly discernible.

12. WETTING ANGLE

The amount of solder used must be enough to form a concave fillet around each component of the joint but not enough to hide the profile of the components beneath the solder. The solder fillet should rise smoothly from the surface of the components to form an angle of not greater than 70° (see Fig. 4.1).

This angle is known as the *setting angle* and may be used to help assess the quality of the joint as follows in Section D.

D – Standard of solder joints

The main points to note are as follows:

Excess solder, providing all other aspects of the joint are satisfactory (i.e. correct tinning, joint formation, etc.) should not constitute a reason for rejection as long as it is confined to one or two joints on any one assembly. However, excess solder on more than one joint in one assembly must always be reported to the production supervision who must investigate the reasons for this with the operator and correct his technique as necessary.

There must be no pin holes, cracks, sharp deep depressions or evidence of inclusion of any foreign matter in the solder surrounding the joint.

There must be no 'spikes' or 'icicles' of solder on any part of a hand-soldered joint. Raising of the solder surface into 'blunted spikes' may occur on certain joints during machine soldering operations and is very difficult to avoid completely. This condition, providing it is not widespread and excessive may be accepted by obtaining the approval of the quality manager.

All components connecting wires, especially thin, varnish-insulated wire which has to be cleaned of varnish and tinned before soldering into a circuit (e.g. 'lead-out' wires of small coils and transformers

mounted directly on to PCBs) must always be arranged at the soldered joint so that the wire can be seen to project from at least one side of the solder in the joint in order that the 'wetting angle' of the solder on the wire may be checked to ensure that the wire was truly cleaned and properly tinned and that no subsequent 'dewetting' has taken place.

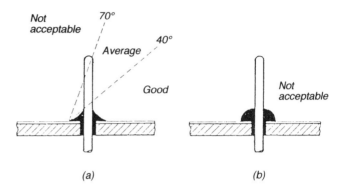

NOTE:
(a) Below 40° – good
     40° to 70° – average
     Above 70° – not acceptable (excess solder and/or poor tinning
(b) Not acceptable – excess solder

Fig. 4.1 Wetting angle.

E – Flat solder tags

1. GENERAL

Figs. 4.2 and 4.3 illustrate the standard methods to be employed for forming wires prior to soldering, when flat tags form part of the joint. Fig. 4.4 shows an alternative for braided conductors.

No more than three separate conductors must be soldered to one short flat tag, but either or both joint forms may be used on one tag. If it is required to connect more than three conductors to one point then special long tags, double-ended or star tags must be substituted for the normal short tag.

2. BRAIDED CONDUCTOR

If a braided conductor (e.g. screened cable) is to be soldered to a flat tag it shall be stretched lengthwise, flattened and tinned, taking care that the solder does not flow by capillary action more than a distance equal to half the width of the braid past the termination of the joint as in Fig. 4.4 then formed to shape and the joint made as in Fig. 4.3. An alternative method of soldering braided conductor to a flat solder tag is shown in Fig. 4.4. Braid is stretched or compressed to fit snugly over tag and is not pretinned.

3. SLEEVES

When a joint is to be sleeved with a rubber or synthetic sleeve, the joint form shown in Fig. 4.2 shall be adopted. The sleeve should not be pushed down over the joint until the joint has been inspected and found satisfactory.

4. WIRE REMOVAL

All forms of joint shown in Figs. 4.2 and 4.3 must be made in such a way that the conductor may be removed from the tag without damage and without undue disturbance to any other joint made to the same tag.

5. LARGE TERMINAL

In the case of large connecting terminals (e.g. large rotary switches) where the heat necessary to make the solder run on the large tag may cause damage to the wire insulation, it is permissible to attach a smaller solder tag to the large component terminal with a screw, crinkle washer and nut and then to solder the conductor to the smaller tag as stated above. This type of joint must be limited to two small tags per large terminal. One small tag to be mounted on either side of the main terminal. Alternatively, two double-ended tags may be used giving four soldering joints.

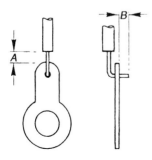

Fig. 4.2 Single-wire conductors.

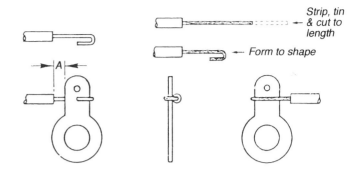

Fig. 4.3 Single- and Multistrand conductors.

NOTES TO FIGS. 4.2 AND 4.3
Dimension A – Not greater than 1.5 mm but long enough to prevent burning of insulation and inclusion of insulation in the solder of the joint.
Dimension B – This must be not more than 1.5 mm or twice the conductor diameter, whichever is the greater, and not less than 0.75 mm.

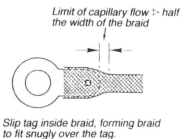

Fig. 4.4 Braided conductor – alternative method.

F – Solder lugs (flat, spade & ring terminals)

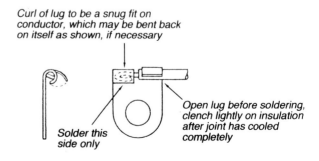

Fig. 4.5 Multistrand conductors.

NOTES:

■ For multistrand conductors with insulation no more than 3 mm in overall diameter

■ For single-core conductors and all conductors with insulation over 3 mm in diameter; the insulation must be stripped so that the conductor projects into both lugs and the insulation terminates between 0.75 mm and 1.5 mm away from wire entry side

■ Adjust both lugs to be a snug fit on conductor which must visibly project through both lugs

■ Solder must fill both lugs

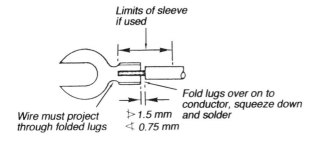

Limits of sleeve if used

Wire must project through folded lugs

Fold lugs over on to conductor, squeeze down and solder

> 1.5 mm
< 0.75 mm

NOTE: solder must run right through folded lugs. Joint may be sleeved with 19 mm long sleeve.

Fig. 4.6 Single or multistrand conductors.

G – Pins & tubular cross-section terminals

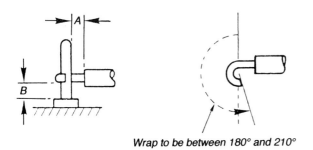

Wrap to be between 180° and 210°

Fig. 4.7 Pin terminals.

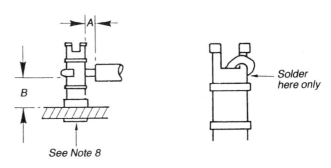

See Note 8

Solder here only

Fig. 4.8 Tubular terminals (turret).

NOTES to Figs. 4.7 and 4.8

■ Dimension 'A' to be no greater than 1.5 mm but long enough to prevent burning of insulation or inclusion of insulation in solder

■ Dimension 'B' to be no more than approximately half the terminal height for single conductors. Where more than one conductor – not less than 1.5 mm for the lowest conductor

■ Maximum of three conductors to one end of each turret terminal. Component wires to be mounted above permanent link wire which connects terminals on the same board

■ When terminating at a hollow turret terminal from the underside, pass conductor up through the turret and bend downwards at right-angles into slot then wrap 180° around top groove

■ Solder to be applied to outside of turret – do not fill bore with solder

■ Only one wire of the permitted three to be terminated in this way on each turret terminal

■ When a turret lug or other similar connecting pin is fitted to a PCB it must always be soldered to the track or pad to which it is intended it should make electrical connection. Solder must tin to both track and turret/pin completely around on both sides of the board. Because of the degree of oxidisation, soldering must be completed within 48 h of fitting

H – Thimble terminals

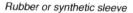

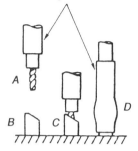

Rubber or synthetic sleeve

KEY

A – Tin wire and cut to length which will just allow insulation to clear top of thimble

B – Tin inside surface of thimble. Do not fill more than half full with solder

C – Maintain solder in thimble molten with iron on outside surface. Insert wire fully into thimble

D – Push sleeve fully down over thimble after inspection

Fig. 4.9 Thimbles on board.

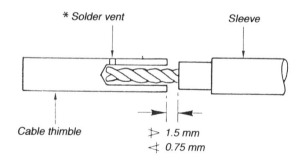

*See Note 4

Fig. 4.10 Thimbles on cables.

NOTES to Figs. 4.9 and 4.10

■ Space between insulation and thimble to be kept to a minimum consistent with no burning of insulation

■ Solder must completely fill thimble and be visible as a concave surface between edge of thimble and conductor

■ Solder should exhibit the same curve and wetting angle as indicated in Fig. 4.1

■ If a solder vent is not provided, care must be taken that no air/flux is trapped inside the blind hole and that the solder completely fills the space around the wire.

J – Thin component leads – acceptable method

NOTE: if the wire is less than half the diameter of the hole in which it is to be fitted then the wire should be cropped and bent right down onto the board on the copper side so that the short end of the wire lies along the copper track. Such joint formation will be acceptable provided that the solder has tinned correctly to the wire where it lies along the copper track even if the hole on the opposite side to where the wire is bent down onto the track has not filled with solder.

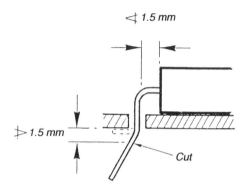

Fig. 4.11 Thin component leads – acceptable methods.

K – Tinning

1. TINNING

Tinning is the act of coating metallic terminations with a film of tin or tin/alloy, it prolongs the storage life by reducing oxidisation of the base metal, reducing interaction between the base metal and the insulation and improves solderability. Particular care should be taken with equipment designed to be in use for long periods (more than 10 years), as bare copper wire can interact with some types of insulation, causing them to become brittle. If in doubt, consult the cable manufacturer.

2. TINNING OF CONDUCTORS

A solder pot is installed containing solder of 60/40 tin/lead solder, of type F, G or K as specified in BS 219 'Soft solders'. A continuous check should be kept on the temperature of the bath, this should be controlled by a calibrated thermostat.

3. TINNING OF COMPONENTS, LEADS AND TAG, PRIOR TO SOLDERING

Solder bath method

The solder bath should be large enough to ensure that the temperature of the solder remains constant while in use (270 ± 10°C). The exposed area of the surface of the solder should be reduced as far as possible by the use of a sheet of asbestos to prevent the components being heated by direct radiation from the bath. The surface of the bath should be kept clean and bright.

The component termination should be immersed in the solder in the direction of its longitudinal axis for about 2 s. The wire termination should be immersed up to 6 mm from the point where the termination emerges from the beds, and the soldering tags should be immersed up to a point 3 mm beyond the place intended for the connection of wires, or half their length if this latter would result in too small a depth of immersion. Good tinning is shown by free flowing of the solder, with wetting of the termination. If this is not found, redipping should not be undertaken without mechanical cleaning, because if the termination cannot be wetted in the first application, further dipping would cause deterioration of the materials.

4. REDUCING DAMAGE

To make connections quickly and reliably, some users insist on using solders containing 60–70% tin, giving a melting point below 200°C, as this allows a soldering temperature to be employed which would reduce the likelihood of damaging components by overheating.

In certain specialised applications, however, where strength at high temperature is required, the tin content may be over 90% with addition of, for example, 2.5% of silver or 5% of antimony. The solders with the highest liquid temperature (around 300°C) and greatest mechanical strength are lead-based with silver and only 40% tin, and in con-

sequence have somewhat inferior wetting properties.

Modern developments include the addition of a small percentage of copper to tin/lead solder, to reduce the erosion of soldering iron bits by the molten solder, the bits may thus be increased by a factor as high as 100, and it appears to have been more successful than the development of new alloys or protective coating for solder bits.

5. DIFFUSION BARRIER

Tin coatings are the traditional finish for any component required to be readily soldered, and many million tags, terminal posts, etc., are coated each year with pure tin or with a tin/lead alloy rich in tin, either by hot-dipping and centrifuging or by barrel electroplating.

The majority of these parts are manufactured from brass and this has caused some difficulties, since zinc will quite easily diffuse out of the brass through the surface coating, which makes soldering difficult especially on parts which have an insufficient thick coating and have been stored for long periods under poor conditions.

A commonly used procedure is to apply a layer of electrodeposited nickel or copper as a diffusion barrier to the brass parts, prior to the final tin or tin alloy coating. In hot-dipped coating of brass tags, etc., a nickel undercoat prevents zinc from passing out of the brass into the dipping bath, where its presence would give rise to coatings with poor appearance and storage properties.

In the recommended procedure, the parts are initially coated with an undercoat layer of 0.0025–0.0050 mm of copper/nickel and are then plated with a pure tin coating not less than 0.0075 mm thick.

6. TINNING PRIOR TO SOLDERING

The two stages in making a soldered joint are:

■ Tinning the metal surface
■ Filling the space between the tinned surfaces with solder

Whether or not these two operations are carried out separately in practice depends on circumstances but, in general, it is advantageous to do so, because each can then be done under the most suitable conditions. The soldering of pretinned parts is usually more thorough, and economical of the operator's time.

7. HOT DIP TINNING

Although the words imply the use of tin alone it is usual for the tinning to be done with eutectic solder, the solder (or tin) may be applied by means of copper soldering bit after treating the surface with a suitable flux. In this case the work should be held so that the solder draws back onto the copper bit leaving a thin flat film that will not interfere with any assembling that a component may require. A method commonly used for small parts such as tags or wire ends is to immerse them one at a time in solder contained in an electrically heated iron pot of the smallest size, preferably well insulated against heat loss.

8. ELECTRO TINNING

Good adhesion of electrotin coatings is essential since loosely adhered coatings are immediately dissolved off without the basis metal being wetted by the solder. The stannous chloride bath which, until a few years ago was the only bath described in instructions on electroplating often produces loosely adherent deposits of uncertain thickness and should not be used as a preparation for soldering. The thickness of the electrotin deposit is a decisive factor for rapid and dependable soldering, and difficulties have often been traced to deposits being too thin. Thin coatings solder quite well for a few days but after standing for a week or two the adhesion is uncertain, due to the diffusion of the tin into the base metal and due also to unseen oxidation. The remedy is to apply adequate thickness of tin, in the case of a brass base a minimum of 0.0125 mm and for a copper base 0.008 mm. Electro tinning effects a small economy in tin as compared with the hot-tinning method of preparation, but the thicker coating produced by the latter method should make it possible to use less solder when making the soldered joints. There is also a saving of time in making the joints when a hot-dipped coating is employed as compared with an electrotinned coating.

9. HOT ROLLER TINNING

Tin/lead alloy coatings produced by 'roller tinning' have reasonable initial solderability, the solder spread being consistent with the relatively small thickness of the coating. After normal storage or long-term damp-heat treatment, however, the area of spread of the solder is usually one-half the original value and the wetting time significantly increases. For the more exacting soldering application this deterioration in solderability would be inadmissible, but the solderability of this type of finish when freshly applied is such that roller coating with tin/lead alloy is commercially rather widely used for printed circuits.

10. SOFT SOLDERING OF WIRE TERMINATION

Care should be taken to see that heat controlled devices or periodically inspected soldering irons are used. The temperature of the bits should not be less than 40°C above the melting point of the solder.

When soldering, the greatest care must be taken to avoid dry joints. It is important that there is no movement at the joint either when soldering or whilst the solder is solidifying. Solder must run smoothly onto both items being joined. Bead formation is usually indicative of a dry joint. If possible, guards should be used to protect wiring or components from damage by the soldering iron. Consideration must be given to the use of thermal shunts, particularly where the length of the lead from a component to its mounting tag is of the order of 6 mm. Thermal shunts should always be used when soldering small rectifiers, transistors, high stability resistors and other components which are particularly vulnerable to heat. Where silver-plated terminators are to be soldered, such as silver-plated terminals or silver-plated PTFE wires, etc., solder with 2% silver content should be used.

All flux residue should be removed from the joints using a suitable solvent cleaner such as isopropyl alcohol. NOTE: trichloroethylene should not be used. It is very important that the flux residue is not washed into the sleeving. Double soldering or touching up of joints should not be permitted, unless some change in the mechanical configuration is made before soldering for the second time.

11. CONNECTING WIRING AND COMPONENTS

Connecting tags and spills without holes should be tinned before assembly. A hot iron and solder should be applied to the termination and should be in contact for about 10 s. The solder should be applied for the first 2 s and tinning as evidenced by free flowing of the solder with proper wetting of the termination should be completed within the first 2 s.

When connecting to spills and tags with holes the above procedure should be followed, except that the end of the wire should be bent at right angles (i.e. L-shaped) and then hooked through the hole in the tag or spill. The wire should not be clenched or in any way further bent so as to hinder its subsequent disconnection.

When connecting the turret tags, the same procedure as before should be followed, except that when a lead is brought through the centre of the tag the lead should be bent along the tag channel.

Semiconductors, diodes, transistors and other heat sensitive devices must be treated with care when soldering. Wire leads for these components should be tinned to within 6 mm of the body or to half the length of the tag of the termination, as appropriate. This means that satisfactory soldering processes cannot be carried out nearer than 6 mm to the body of the device, and a heat shunt between the point of soldering and the body of the device should always be used. To avoid damage to components which may be caused by high 'g' shocks when cutting wire leads with side cutters (with wedge action), leads should be cut with a shearing action, as shocks so generated can cause complete breakdown or intermittent faults within the components.

Where it is necessary to connect more than one wire, the wires should be individually connected and not twisted together before soldering, and care should be taken to see that the ends of the wire termination hooked around tags or spills, or projecting through holes in the item are not long enough to cause damage by shorting or brushing. Excess wire should be cut before soldering, and the soldered joint should be kept as small as possible with a smooth surface, free from bumps and tails.

12. STRIPPING INSULATED WIRE

Damage to conductors must be avoided, particularly on single strand or small diameter PVC-or polytetrafluoroethylene (PTFE) insulated wires. Stripping of insulation from covered wires should only be carried out with proper mechanically covered wires. Special mechanical stripping can be obtained for PTFE-covered wires.

L – Inspection/ supervisor points to note

1. EFFICIENCY OF JOINT FORMATION

Two types of defective joint are considered and formal identifying names have been give, they are currently termed 'misses' and 'partials'. Poor joints may be formed when dirt or oxides are present on the components. Oxides are the result of porous plating which allows either copper oxides or sulphides to migrate through to the surface and prevent acceptable fillet formation. Double soldering to fill voids left from the original soldering is ineffective, the second application melts the first fillet but does not increase the final fillet size and disperses the copper/tin alloy plane with a corresponding joint strength decrease, but this is not true if some change in the mechanical configuration is made between solderings.

Since the strength of a soldered joint is essentially the shear strength of the solder in the fillet, the following joint strengthening methods have been tried with and without success: notching, crimping, flattening, flaring or otherwise altering that portion of the wire which is included in the solder fillet, using eyelets in plated-through holes on printed circuits or using another material on the back of the board around the leads or controlling the length of lead which is in the solder.

Of the methods described above, flaring of a position of the lead which is included in the solder fillet is the only method which shows significant strength increase. Flaring is accomplished with a tool which cuts and flattens the wire so its greater dimension exceeds the hole diameter in the board. In this way not only the shear strength of the solder but also the interference of the flattened portion of the wire contributes to the overall joint strength.

CAUTION: When brass eyelets are staked into the holes in the PCB and when the eyelets are unplated or thinly plated, the joint could deteriorate due to zinc contamination.

Plated-through joints, on examination, reveal that there is a significant tendency for voids to form.

2. MARKING OF JOINTS AT INSPECTION

Coloured lacquer or inspection paints which have a resin base should not be used, especially where moisture and temperature can produce low insulation, or if the assembly is to be encapsulated.

NOTE: The marking of soldered joints is permitted in some organisations. In such circumstances, adhesive 'arrows' normally indicate a joint that requires reworking.

3. INTERMITTENT FAULTS

Intermittent faults are sometimes the most difficult to locate and diagnose. They generally fall into two broad classes:

(1)   Faults usually arising under low voltage conditions which rectify themselves abruptly when the voltage (or available energy) is raised
(2)   Faults which come on and off in phase with some operating condition, e.g. temperature, humidity, voltage or vibration

Disconnections such as those described in (1), are nearly always due to a dry joint at which an oxide film slowly builds up, but which can be broken down by increased voltage. Short-circuit or low insulation faults in this class are due to metal whiskers or silver/iron migration or occur singly due to corrosion.

With faults described in (2), above, diagnosis can be related to operating conditions, and measurements can be made when the assemblies are successively warmed, cooled and vibrated.

4. SILVER ABSORPTION

Open circuit and high resistance faults have occurred on many silver-plated component terminations owing to the use of normal 60/40 solder, and it is a fact that at least 2% of the silver plating is absorbed into the solder during soldering to leave unwetted surfaces.

The melting temperature of the silver on a plated termination would be 179°C allowing a bit temperature of 219°C to be used, and the average melting temperature of 60/40 solder is 188°C with an allowable bit temperature of 218°C, i.e. 30°C in excess of liquids. The usual bit temperature is in the range 220°–250°C. The thickness of the silver plating does not improve solderability and the poor area of solder remains constant, due to the fact that the solder is only wetting a doubtful silver surface.

5. SOLDERING PLATED SURFACES

Cadmium plating

This type of plating is used as a protective coating on steel, copper and brass components of radio and telephone apparatus. Its soldering characteristics are decidedly superior to zinc, using the resin-in-alcohol type of flux, but under practical conditions operators find that soldering on cadmium requires a hotter iron than on tin, or they must apply the iron for a longer time; the rate of output of sound joints is slower on cadmium-plated than on tinned surfaces.

Silver plating

Silver is resistant to oxidation and is sometimes used for electroplating brass and copper components. It can be soldered readily with the usual quantities of tin/lead solder using noncorrosive resin-in-alcohol fluxes. Cases of difficulty have been traced to the silver coating being much too thin and, therefore, being rapidly dissolved from the surface by the molten solder. For certainty the silver coating should be no less than 0.00075 mm, preferably 0.0125 mm, and a solder containing 2% silver used to make the joint.

6. SOLDERABILITY

Since the soldering fluxes normally used in printed circuit manufacture are not designed to remove organic contaminants such as grease, wax and oil from the surfaces to be soldered, these foreign materials should be removed with organic solvents since they can seriously affect the surface solderability. In this way, some assurance can be obtained that the soldering flux will come into intimate contact with the metallic surface and perform the prescribed functions instead of meeting with an organic interference layer which inhibits the flux from properly cleaning the metallic surface.

The flux must also have sufficient chemical activity to react with metallic tarnishes in order to provide the clean base metal surface required for good wetting.

In these cases where the metallic tarnish on the

surface of the base metal is either too heavy for complete chemical removal, or too stable for the fluxing reaction to be effective, precleaning the surfaces with a chemical solution that will remove most of the metallic tarnishes is mandatory. In this way, solderability is established before, and confirmed when, the selected flux is used.

7. SOLDERABILITY TESTING

Solderability specifications can easily be incorporated into the purchasing instructions given to the vendors. They can also be used as quality control methods designed to assure reliable soldered connections during the manufacturing process.

Since the surface to be soldered can deteriorate after handling and storage, it is essential that additional solderability checks are made after the incoming inspection. The most advantageous time for this preproduction solderability check to take place would be immediately prior to soldering.

The solderability test that takes place immediately prior to soldering frequently involves the use of the actual production line flux, and the soldering time and temperature duplicate production conditions.

Bear in mind that PCBs or components found unsatisfactory during the incoming inspection can be rejected and returned to the vendor, whereas parts found unsatisfactory during the preproduction solderability test must be subjected to a solderability restoration process. For this reason, it is important to couple pre-production solderability testing with a well formulated procedure for restoring solderability. In this way, a group of components or PCBs exhibiting poor solderability can be forwarded to a restoration area and corrected prior to assembly and final soldering. There is no reason to anticipate that a board or component that is not solderable prior to its incorporation in the production process will be solderable during the production process. In addition, it is also far less costly to reject components or boards and return them to the vendor, or restore the solderability of unsolderable components or PCBs prior to assembly, than to attempt to repair poor solder joints after the soldering operation.

Note that a lack of surface solderability which manifests itself in poor wetting, in most cases, cannot be corrected by reheating or touching up the solder joint. When dealing with repairs it should be remembered that the soldered joints are important and that the rectified joint must be subject to the same parameters that control the production process.

8. SOME SOLDERING PROBLEMS AND THEIR SOLUTIONS

Burning PVC wire

Product design has been changed to include the use of relatively inexpensive PVC-insulated connecting wire. This type of insulation has a melting point of 148°C vs. the 345°C or higher melting point of previously used materials. Because of the low melting point, operators find it very difficult to make solder connections without burning the PVC insulation. The heat involved in making the connection results in insulation char and bloom. Previous studies indicated that a low temperature solder should be used and that a proper iron was essential. The iron should be capable of maintaining constant heat output during the entire soldering operation.

Economy in these operations dictates the use of solder alloy having a low tin content (40/60), which unfortunately has a melting temperature of 234°C. By changing this alloy to 60/40 the melting temperature is lowered to 188°C, a difference of approximately 46°C. The temperature difference between these alloys reduces the dwell time of the soldering iron on the connection, making the soldering of PVC wire more feasible. The combination of proper solder alloy and soldering iron results in satisfactory use of PVC-insulated wire at substantial savings.

Soldering fine wire

Many soldering problems have been encountered in electronic equipment using copper wire as fine as human hair. The use of this wire has increased greatly in recent years with the development of low power circuitry and microcircuit devices.

This small diameter copper wire is generally insulated with a polyurethane-based coating containing various additives which improve specific physical characteristics. Nylon is commonly added to improve flexibility and reduce surface friction. The basic advantage in using polyurethane-type formulations is that these thermo-plastic materials flow away from the surface of the wire upon the application of proper heat during the soldering operation without leaving a residue. Metallurgical and mechanical problems, however, plague the user of this wire. The tin in the solder dissolves copper at soldering temperatures and can embrittle the wire beyond its useful strength. The wire can easily be stressed mechanically causing either nicking or size reduction.

Companies engaged in manufacturing memory products are large users of fine copper wire and are continually trying to improve their manufacturing and soldering techniques. Close parts, proximity and heat-sensitive materials complicate the efforts made to promote product reliability. However, by controlling the soldering temperature, solder alloy composition, and type of flux used, and by establishing the proper soldering procedures, reliable soldering connections can be made. The elimination of temperature variances minimises the number of rejects involving nicking down of wire, embrittlement, lifted pads, delaminated and measled boards, and (probably the largest single problem of all) tunnel connections resulting in an intermittent and ultimate connection failure. A tunnel connection results when the nylon/polyurethane coating on the wire is not completely dissolved during the soldering operation. Proper tool and materials selection, as well as improved soldering techniques, make the soldering of fine copper wire feasible.

Flexible circuits

The use of flexible circuits instead of hand-terminated individual wires has permitted large savings in repair work since miswires are eliminated. The circuitry achieves the high density, low weight packaging so important to modern equipment. The flexible circuit has, however, created some special soldering problems. Since it is basically a printed wiring board on flexible material, all the regular printed wiring problems plus some new ones have been encountered. Since the basic flexible circuit was very thin (0.2 mm) and pad sizes were very small, cored eutectic solder (flux centre) of a small diameter size was necessary. These small-core solders have contributed to tip dewetting problems which can be solved by flushing the tip with 1.58 mm diameter solder. A microminiature soldering iron having good heat retention properties, with a working range of 274°–288°C, is satisfactory for this job.

9. ELECTRONIC COMPONENTS ASSEMBLY

Mounting of components

In the case of wire-wound components, the length of the lead from the component to its mounting tag should be less than 6 mm. For wire-ended components less than 14 g in weight and other components

that have no method of mounting, the leads should be soldered to rigid pillars, the distance from the joints of emergence of the lead to the pillar being 6–12 mm. Components having radial leads or unusual mass distribution will need special mounting, and wire-ended components weighing more than 14 g should be rigidly clamped. Special components which are not normally clamped should be cemented with a little epoxy resin or a similar adhesive.

Wiring: cable forms and leads

The edges of all clamps and clips should be rounded. Care should be exercised in the running of insulated wire in that it is not carried over, or bent around, any sharp corner or edge that might in time cut through the insulation. Leads shall be of such a length that they are not unduly tight between points of connection and are not stressed. Sharp bends should be avoided with PVC or similar covered wire, as otherwise the conductor may in time force its way through the insulation. Sharp bends must be avoided in extra high-tension wiring where they might give rise to brushing. Wires crossing hinges should be looped to avoid stress on the conductors.

Cables and screened leads

When fitting cables, particularly metal braided cables, sharp bends must be avoided. Flat twin cables should be bent on a minor axis. Cables should not be carried over or bent around any sharp corners or edges otherwise damage may occur, and where a cable is carried across a hinged portion of a chassis or brought out to form connection with instruments or apparatus, especially if these are flexibly mounted, sufficient slack must be left to prevent stress on the conductors. The ends of metal braiding or screened cables must be finished off neatly, loose strands or ragged ends should not be permitted. Cables must not be stretched too tightly between successive clamps, as this tends to reduce the overall diameter of the cable and may result in weakening the insulation.

Sufficient slack should be left at the ends of cable forms to allow the removal of components to which they are connected for inspection or maintenance.

M – Check lists

1. FINAL INSPECTION SOLDERING

Check that:

- There are no unsoldered joints
- There are no dry joints (where more than one wire is connected to the same point, care must be taken that all wires are soldered)
- There is no excess solder on joints (lead form must be seen through the solder)
- Cold or crystalline joints do not exist
- There is no flux residue
- There are no tails of wire or spikes of solder
- Where leads are taken to lugs or pins of round form, the wrap-round is between 150°–220°. On pins of less than 2 mm 360° wrap is permissible
- Where leads are taken through holes (such as PCBs) the wire should be bent at right angles (i.e. L-shaped)
- Where wires are passed up through turret lugs, the wire is bent at right angles (i.e. L-shaped) into the channel. Wires will not be passed down turret lugs
- When soldering wires to plus and sockets, the solder bucket is filled with solder and it is ascertained that all spikes and beads of solder are removed
- Where heat sensitive components are used (i.e. diodes, transistors, etc.) these are fitted last. Ascertain that heat sinks are being used
- There are no lifted tracks on PCBs
- PCBs are free from burns and damage
- There are no short-circuits between tracks
- There are no blow holes in soldered joints

2. SUPERVISOR'S CHECK LIST

It is the responsibility of the line supervisor to check that:

- The operator is using the correct iron for that particular operation
- The body of the iron is undamaged and the cable and plug are in a satisfactory condition
- The face of the tip is flat, at the correct angle (i.e. short taper) is free from pits, hollows, and is clean and tinned

- The tip temperature is at the recommended 288°C using an Adcola iron/constantin thermocouple loop or similar calibrated thermocouple
- The solder being used is the correct type
- All flux residue is removed, using isopropyl alcohol proprietary flux cleaner

3. DESIGN REVIEW CHECK LIST

- Does the unit require special handling?
- What special techniques are required in the repair or replacement of components?
- Are the components having the highest failure rates readily accessible for replacement?
- Is soldering adequately specified? What provisions have been made to prevent cold joints and to ensure removal of flux?
- Is the design such as to minimise soldering iron burns during manufacture and maintenance?
- Have the standard materials been specified in all possible cases?
- Does a heat-dissipation problem exist on PCBs?
- Are transistor, diode and tantalytic capacitors properly polarised on PCBs?
- Have all specifications been met unconditionally?
- Does any specification require modification?
- Can any reasonable or unusually difficult requirement be relaxed?
- Does the design provide for adequate protection of maintenance and test personnel against accidental injury?

N – Examples of soldering

1. CARE OF TOOLS

The main points in maintaining soldering tools are as follows:

Soldering iron

- Correct bit must be used for the job in hand
- Bit to be clean *at all times*
- Iron must be temperature checked on a regular basis

- Bit must be rotated regularly to ensure correct contact
- Bit must be replaced when uneven or pitted

Sponge

- Sponge must be damp prior to soldering
- Sponge to be replaced when it is badly worn and can no longer remain damp during soldering

Cutters and pliers

- Cutters and pliers must be maintained in good working condition at all times

IMPORTANT NOTE: Unnecessary handling of PCBs must be avoided. Handle by the edges only, or use cotton or linen gloves to prevent contamination from perspiration.

2. WIRES OUTSIDE TERMINALS

Any strand which is unattached to a terminal is a Class C defect. Where an unattached strand is likely to cause a short-circuit, it is a Class B defect.

Where an unattached strand is causing a short, or more than 10% of the strands are free, this is a Class A defect.

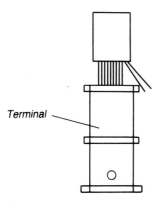

Fig. 4.12 Wires outside terminals.

3. EXCESS OR INSUFFICIENT SOLDER

Too little solder can cause a high resistance joint. Too much solder can hide dry joints.

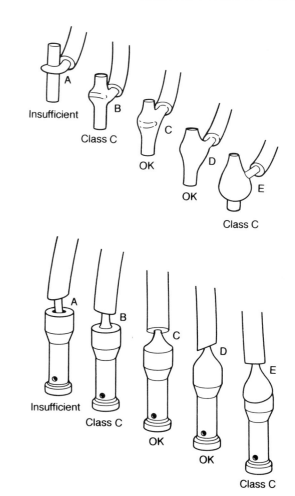

Fig. 4.13 Excess or insufficient solder.

4. DAMAGED INSULATION

Insulation shrinkage and drainage caused by too much heat.

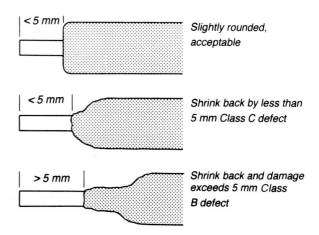

Fig. 4.14 Damaged insulation.

5. UNSOLDERED JOINTS

An unsoldered joint is a joint completely devoid of solder. Any joint that has no solder is a Class A defect.

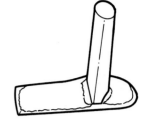

Fig. 4.17.

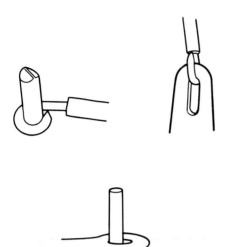

Fig. 4.15 Unsoldered joints.

6. DRY JOINTS

Terminals

Dry joints are distinguished by having a distinct line between the solder fillet and either the terminal or the component lead.

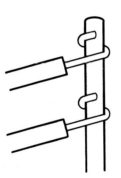

Fig. 4.16.

Dry joints on PCBs also show a line between the lead and the solder. Joints will often show an angle between the solder and the lead that reflects the reluctance of the solder to wet the lead.

7. BRIDGING

Low soldering iron capacity or too much solder can lead to bridging between pads or leads.

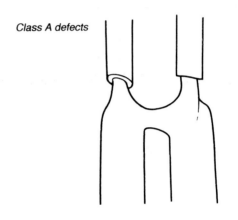

Class A defects

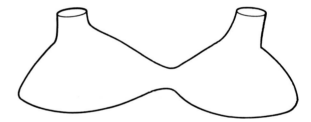

Fig. 4.18 Bridging.

8. SOLDER SPIKES

Spikes on joints are caused by the soldering iron temperature being too low or the iron bit is the wrong size.

- Spikes up to 50% of joint minor diameter – no defect
- Spokes up to 100% of joint minor diameter – Class C defects
- In conditions of high packing density such as multiway and socket tags, spikes over 25% of joint diameter are Class B defects

■ Spikes over 100% of joint minor diameter or spikes of any length which are liable to cause a short-circuit are Class B defects

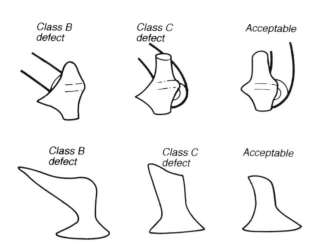

Fig. 4.19 Solder spikes.

9. JOINT HEIGHT

Joint height should not be less than 1 mm and no greater than 3.2 mm.

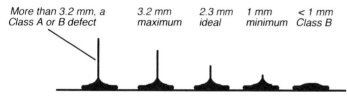

Fig. 4.20 Joint height.

10. EXCESS OR INSUFFICIENT SOLDER

It is essential to use the correct amount of solder. Too much may hide a dry joint. Too little will provide a high-resistance joint that will overheat or fail.

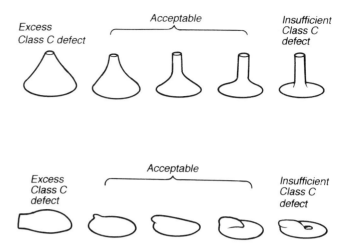

Fig. 4.21 Excess or insufficient solder.

11. HOLES IN JOINTS

There should be no holes in the joints. These arise when the joint is formed without allowing the solder to flow completely around the component lead.

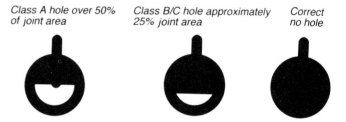

Fig. 4.22 Holes in joints.

12. LIFTED LANDS

Excess heat applied to the joints will lift the track or pad.

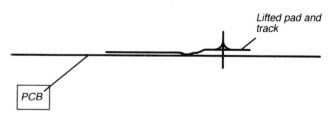

Fig. 4.23 Lifted lands.

13. GUIDE TO FAULTS – FLOW SOLDERING

Guide to faults

Flow soldering quality depends upon the PCB design and the setting-up of the flow solder machine.

Assuming that the PCB design is correct, the following points are a guide to determine the reason for soldering problems.

14. SOLDER BATH CONTAMINATION

The following guidelines are to be used to ensure correct chemical content of the solder bath. Flow solder alloy used in the solder bath is to be 63/37 tin/lead nominal by weight with a tolerance of 61–65% tin and lead to balance. Should the tin content fall below 61%, new solder should be added to the bath and it should then be re-analysed. If the following contamination limits are exceeded the bath should not be used:

Copper	0.35% max
Antimony	0.40% max
Gold	0.35% max
Arsenic	0.05% max
Bismuth	0.10% max
Cadmium	0.003% max
Zinc	0.002% max
Aluminium	0.002% max

A label will be affixed to the machine showing the 'next date of solder check'.

15. FLUX

The specific gravity (SG) of flux must be checked prior to use and topped up with new flux if found out of limits.

When the flow solder machine is used all day the flux SG is to be checked a minimum of once in the morning and once in the afternoon. A record card will be displayed on the machine showing dates and times of the flux checks.

Table 4.1 Solder bath problems/solutions.

Conditions	Cause	Remedy
Dewetting	Oxidation through plating	Clean surface chemically
	Exposed copper	Clean surface chemically
	Porosity of gold (electroplate)	Clean surface chemically
Icicles	Dewetting as above	Clean surface chemically
	Temperature too low	Raise temperature
	Speed too fast	Decrease speed
	Immersion too deep	Raise conveyor
	Not close enough to wave adapter	Lower wave and adjust conveyor
Bridging	Contamination	Chemical clean surface
	Dust and dirt accumulation	Solvent cleaning
Blow holes	Hole gap too large (conductor to pad)	Redesign with smaller holes-speed up conveyor
	Flat lead–round hole	Redesign flat hole
	Gold plate on leads	Tin coat lead – lower temperature
	Dirty lead	Clean
	Burrs in hole	Scrap
	Coating on components melting in hole when flush mounted	Raise components above board level e.g. by crimping lead
Dirty solder application	Solder contaminated calcium, zinc, gold, etc.	Check solder, if bad, scrap
	Gold scavenging	Increase speed, lower temperature
Valley in plated-through holes	Solder too hot	Lower temperature
	Draining down lead	Increase speed, lower temperature
	Temperature too low	Raise solder and preheat temperature
	Gap too large	Redesign PCB layout, use thicker lead components

16. INSPECTION NOTES

There shall be no blistering or delamination of the PCBs, neither shall there be any lifting of tracks or pads. Solder fillets shall be smooth, shiny and free from fracture; in addition there shall be no solder splashes or bridging between adjacent areas. NB: a solder splash may not be shorting, yet when large enough can become dislodged from the laminate causing a short circuit.

17. DEFECTS

Solder splashes causing a short circuit – Class A.
Solder splashes likely to cause a short circuit – Class B.

Solder splashes likely to cause a small short circuit – Class C.

18. QUALITY

General quality and form of solder joints as illustrated under Subsection 4.N Examples of soldering, is still applicable.

SECTION 5:
CLEANING ELECTRICAL CIRCUITRY

A – Cleaning solvents

When it becomes necessary to clean electrical assemblies such as PCBs, switches, relays, etc., solvents often have to be used as the only practical method of removing the soiling.

Such solvents must always be used with great care as the more effective the solvent is in removing soiling, the more likely it is to have a detrimental effect on some plastics (e.g. polystyrene capacitor mouldings), coding inks, insulating sleeving, etc.

Certain types of proprietary switch cleaner liquids are known to dissolve the plastics used for certain rotary switch parts. Only use cleaners or solvents known to be safe for cleaning for rotary switches, e.g. RS® solvent cleaner. This cleaner is supplied in an aerosol can. It must be sprayed, with the nozzle provided, directly onto the contacts at close range and in sufficient quantity to wash dirt and old lubricant clear of the switch contacts.

After cleaning a rotary wafer switch with such material it is imperative that, before use, the switch is relubricated by the application of a suitable lubricant such as RS® contact cleaner lubricant supplied in an aerosol, and must be used as directed on the can.

No attempt must ever be made to clean flux residue from soldered connections on rotary switches, relays or any contacting devices by indiscriminate flooding with solvent liquid. The solvent will dissolve the resin and the mixture may then spread over the contact area to form a thin, hard film of resin as the solvent evaporates. Much less harm will be done by leaving small amounts of flux residue on the soldered joints.

After the machine soldering of PCBs, when an oil intermix is used during the flow-solder process, it is essential to remove the flux/oil residue left on the PCB after the operation as the oil can give rise to mould growth.

Under such circumstances polystyrene capacitors must not be fitted to the PCB until after the flow-solder and cleaning operations have been completed. The polystyrene material is known to be very vulnerable both to the heat of the flow-solder operation and to the solvent liquid used to remove the flux/oil residue and these capacitor types must be fitted to the board by hand soldering after the general cleaning operation has been completed.

When it is necessary to scrub the soldered side of a PCB with a brush and solvent liquid to remove traces of resin flux, etc., the board must always be held with the solder side downwards all the while the cleaning is taking place and be kept in this position until the solvent has evaporated and the board is dry. This will prevent the solvent from running onto the component side of the board and causing damage to the majority of the components and materials which are vulnerable to the solvent.

B – Safety aspects of cleaning materials

1.0 INTRODUCTION

There are two aspects to be considered regarding safety aspects of cleaning:

- Safety to personnel
- Safety to the equipment and its components

Main safety points

Most safety points regarding cleaning materials are listed below:

- All cleaning solvents not in use must be stored away from the manufacturing areas – preferably in a fire-proof container/storage room
- All cleaning solvents must be stored in the containers supplied by the solvent manufacturer/supplier
- Solvents must *never* be mixed unless recommended by the supplier
- All containers must be marked to clearly identify their contents e.g. 'contaminated solvent'
- All contaminated solvents must be disposed of correctly. Solvents should not be discharged into the local council drains without official clearance
- Areas where solvents are stored must be clearly identified
- All areas where cleaning solvents are used or stored must be designated '*No smoking*' areas and comply with current fire regulations. If in doubt, seek advice from your health and safety representative or the local fire prevention officer
  NOTE: the storage of machine solder flux falls into this category
- The temperature of the containers must be maintained below the 'flash point' of the cleaning solvents

- All containers used must be in good condition with *secure* lids or tops
- Containers used other than supplied, must be checked to ensure that their material is compatible with the cleaning solvent to be stored
- Operators must take safety precautions as applicable, e.g.
  - Protective clothing should be worn
  - Protective eye goggles/glasses should be worn
  - Protective gloves should be worn
  - Solvent resistant footwear should be worn
  - All containers must be released when not in use

Siting cleaning areas

The siting of cleaning areas should satisfy the following conditions:

- The area must be well ventilated – fume extraction systems should be installed if necessary
- All cleaning areas should either be a restricted area or, at least, not a common thoroughfare
- The area should be as close to the production area as possible
- All benches and floor coverings should be solvent resistant

First aid

At least two members of staff (ideally, one female and one male) should be formally trained in first aid and their training should cover, as a *minimum*:

- Solvent splashing in the eyes
- Solvent splashing on the skin
- Accidental swallowing of cleaning fluids
- Minor burns

For larger companies, the number of first aid personnel should be higher.

Component safety

The compatibility between the components and cleaning fluids should always be checked prior to the full production. It is normal not to immerse or vapour clean any components that are not fully sealed.

Should there be any doubt about a particular component or solvent, it is advisable to immerse samples of suspect components in a sample of the cleaning fluid to be used and check for the following:

- Change in colour of the component body or coating
- Change in the surface of the component, e.g. cracked, tacky, brittle, etc.
- Change in the component's electrical performance
- Component identification affected by solvent
- Any other visible changes

If in doubt check with the component supplier *and* the solvent supplier.

Paint work and plating

The compatibility of paint work, plating, etc., should be checked as above, by immersion of a sample.

SECTION 6:
ADHESIVES & LOCKING
COMPOUNDS

A – General

Adhesives, glues and cements are materials used for uniting other materials so that they adhere more or less permanently. In general, an adhesive should 'wet] the surfaces being bonded and on solidification should not disrupt the bond. The solidified adhesive should not be more rigid than the parts being bonded otherwise an external load will give rise to premature failure of the joint. The adhesive must have adequate tensile strength.

The existence of adhesion is ascribed to inter-molecular forces (i.e. specific adhesion). It is also possible that mechanical adhesion may make a significant contribution to the strength of some joints through the physical interlocking of the adhesive into the pores or irregularities of the surface being joined.

There are many types of adhesives and cements and the most important may be grouped as follows:

- Animal glues, made from the hides and bones of animals
- Fish glue, made by treating fish bones
- Casein glues, made from protein derived from milk curd
- Vegetable adhesives, made from natural gums and starches
- Shellac, a natural resin
- Rubber cement, made from both natural or synthetic rubbers
- Waxes, both natural and synthetic
- Synthetic resin adhesives
- Cyanoacrylate adhesives
- Silicone sealant

Although strong bonds may be obtained with the first four groups of glues listed, being made from animal or vegetable material, they have limited resistance to water and mould growth, and are mainly used for the bonding of wood, paper, leather, cloth, etc. Shellac, a natural resin, is soluble in many solvents and is softened by heat. It was used as an electrical insulating varnish in the past but its use has now been largely replaced by synthetic resin materials.

Synthetic resins and cyanoacrylate adhesives now come in a wide variety, for almost all applications. Care should be taken to ensure that the correct adhesive is used for the job in hand.

Silicone sealant is frequently used as a general purpose glue, and has a high reliance for bonding dissimilar materials such as plastic sheet and metal.

B – Instructions to be observed when using adhesives

The surfaces to which adhesive is to be applied must always be clean and absolutely free from grease or dust. It is always advantageous, where possible and practical to wash or swab surfaces to be bonded with a solvent-type cleaner which will evaporate without leaving a residue immediately before the application of the adhesive.

The adhesive manufacturer's instructions must always be followed when using adhesives of any kind. This is especially necessary with certain of the synthetic resins where the margin between complete success and absolute failure of the joint is rather narrow and entirely dependent on following the manufacturer's instructions precisely.

Care must be taken when applying adhesive, to ensure that only the correct amount is applied and only to the surfaces to be bonded so that the resultant joint is clean and neat. Too much adhesive may also force the incorrect alignment of the pieces to be joined.

It must be remembered that all adhesives take time to 'set' or 'cure' and any movement of the jointed items relative to each other during this period will almost certainly result in an inferior bonded joint.

Some adhesives will only set in the absence of air, they are called anaerobic adhesives. Care must be taken to ensure that all air is excluded from the joint when these adhesives are used.

C – Permanent adhesives

The following is a list of permanent adhesives. The list is not necessarily complete nor is it intended that the existence of such a list shall exclude the use of any other type of adhesive. However, all adhesives used for production purposes must be adequate in all respects for the purpose for which they are intended to the satisfaction of the design authority.

1. RESILIENT ADHESIVES

NOTE: This group includes adhesives which when 'cured' or 'set' retain a more or less 'rubbery' nature and would be appropriate for bonding nonrigid materials such as rubber sealing strip to box lids, or where exact alignment is not a requirement

- Polimide, hot melt glues
- Bostik™ (C and D), a rubber solution for bonding rubber, plastic, metal, etc.
- Evo-stik™, an impact adhesive
- Pliobond 20™, a rubber solution
- RS® silicone dielectric gel, a two-part curing silicone compound used for sealing adjustable coil cores or encapsulation purposes.

2. RIGID/SEMI RIGID ADHESIVES

NOTE: This group includes adhesives which 'cure' or 'set' to a hard, brittle form and would be appropriate for bonding nonflexible materials, i.e. metal to metal, etc. It also contains semi rigid adhesives that will allow joints between flexible materials such as plastics.

- Araldite™ (includes powder and two-tube paste type), an epoxy resin adhesive for bonding most rigid materials including precision jointing of fitted metal parts. Also used as a rigid encapsulant for small electronic assemblies. Has no practical solvent for de-encapsulation
- Polyurethane varnish, for the protection of coils, PCBs, etc.
- Loctite™ 312 (with Activator F), an *anaerobic* adhesive (it will remain liquid while exposed to air but will bond virtually immediately when confined in a joint) can be used to join metals, glass, ceramics and plastics such as polyesters, phenolics and nylon. When aluminium is bonded to steel the handling strength is obtained in 10 s and 60% of full strength in under 2 min, the ultimate tensile strength is 281.3 kg/cm² (4000 PSI).

Warning

Loctite™ anaerobic engineering adhesives can cause dermatitis. In case of skin contact remove immediately by washing with water. Cyanoacrylate adhesives give a virtually immediate, tough, permanent bond. If you get adhesive on your skin do not touch anything

or you will stick to it and this will be serious as the bond will be strong enough to tear your skin if you try to pull away. Wash adhesive off skin with plenty of water as soon as possible. Use only in a well ventilated place.

3. UV CURING ACRYLICS

Ultraviolet (UV) curing acrylics set when exposed to an ultraviolet light source. These adhesives can bond a wide range of substances, typical applications are for bonding small items (glass to metal), PCB conformal coatings and potting of small electrical items such as cinch connectors. A typical adhesive is Loctite™ 330.

D – Locking compounds

1. LOCKING COMPOUNDS

In this context a locking compound is a material which is applied to a screw thread to stiffen the action of the screw in its mating part to prevent accidental movement of the two parts relative to each other.

Two basic types of locking material are involved.

- A semipermanent locking medium which is applied to screws or nuts used for assembly purposes which would not normally have to be removed after initial assembly. The screw thread may be disturbed after locking, but a fresh application of locking compound would then be required to relock the screw thread
- A nonpermanent looking medium which when applied to a screw thread will hold the screw thread in any set position against the action of vibration but will allow the screw to be moved if necessary without having to apply further locking material.

2. SEMIPERMANENT LOCKING MATERIALS

(1) Locking varnish
(2) Loctite™ 221 screwlock, mild strength locking

for all normal locking assembly screws and nuts where a crinkle washer is not appropriate
(3) Loctite™ 241 medium strength locking for cases where (2) would not be a strong enough lock
(4) Loctite™ 270 high strength locking for locking screw studs
(5) Silcoset or Silastic 732 for sealing adjustable coil cores which have been set and require no further adjustment
(6) Tropical wax (Claud Campbell LPRM3) as alternative to Silcoset or Silastic 732

3. NONPERMANENT LOCKING MATERIAL

■ Rocol core locking compound No. 8G a thixatropic material.

E – Approved adhesives & solvents for PCB track repairs

1. GENERAL

The minimum standard required is for the adhesive bond to withstand two soldering operations as described in BS 4025 clause 9e. The ideal adhesive is a one-component rapid curing compound which complies with the above requirement.

Generally, the commercial adhesives available require hot curing for rapid results and their temperatures are usually above the safety limit for an assembled board. Some epoxy adhesives are in solid form and offer some advantages in handling but also require hot curing.

Cold cure adhesives normally require a minimum cure time of 24 h but this can be reduced by the application of some local heat which can also provide a more heat resistant final bond.

It is important that the curing procedure recommended by the adhesive manufacturer is strictly followed, otherwise the resulting strength of bond will be unacceptable. A list of adhesives is given in Subsection C above.

2. PREPARATION OF BOARD

It is essential that all surfaces are clean and should be prepared as follows:

(1) Remove excess solder from any joints involved using a suction tool or other suitable means
(2) Clean the laminate and the track with a suitable solvent using a small brush. It may be necessary in some cases to use a fine abrasive paper (P800) to clean the underside of the copper.

3. APPLICATION AND CURING OF ADHESIVE

Prepare the adhesive according to the manufacturer's instructions; and using a small spatula, apply to both surfaces. Avoid excessive adhesive.

When using a solid film adhesive, a suitable size piece should be placed in position and if desired can be made tacky by the application of a very small quantity of alcohol or acetone. Remove excess adhesive after bringing the two surfaces together. If room temperature curing is to be used, the surfaces must be kept in contact under pressure until the cure is complete.

Curing time may be accelerated using a suitably shaped soldering iron bit in a normal electric iron but the dwell time should not be excessive (normally a few seconds at approximately 150°C – refer to the manufacturer's instructions). Maintain pressure on the repair for the remainder of the cure time.

4. LIQUID ADHESIVES

■ Araldite™ double bubble, for small quantity mixing

5. SOLID ADHESIVES

■ Polyamide hot melt glues

6. SOLVENTS

The following solvents are free from deleterious effects on paper-phenolic and glass-epoxy laminates:

(1) Methylated spirits
(2) Isopropyl alcohol
(3) 1.1.1 Trichloroethane

(4)   Genklene (ICI)
(5)   Chloroethane NU (Dupont)
(6)   Arklone L (ICI)

Caution

Items (1) and (2) are inflammable. In working with solvents care should be taken to minimise contact with the skin and the inhalation of vapour.

F – Lacquers and sealants

In general, lacquers and sealants fall into three categories.

■ Air drying
■ Heat drying
■ Two-part curing

Within the electronics industry there are many lacquers and sealants on the market and particular types suitable for specific requirements must always be agreed between the manufacturer and the supplier with the full approval of the customer.

The following points should be considered when selecting lacquers and sealants:

■ Is the lacquer/sealant necessary or desirable?
■ Is the lacquer/sealant compatible with the equipment temperature and humidity specification?
■ Will the sealant used withstand the equipment shock and vibration limits – in particular, sealant used on electrical adjustments, e.g. potentiometer?
■ Can the lacquer/sealant be applied by brush or spray? If the sealant is in the form of a conformal coating, damage to components due to immersion should be checked
■ Does the colour/finish of the lacquer/sealant detract from the quality of the finished product?
■ Can rework/repair work be carried out once the lacquer/sealant has been applied?

SECTION 7: REWORK & REPAIR OF PCBs

(Bob Willis, Dimension 2 Technology, Reading)

A – Rework – component replacement

1. COMPONENTS WITH AXIAL OR FLYING LEADS

Method 1 (To prevent track damage)

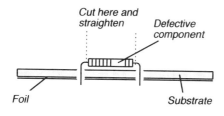

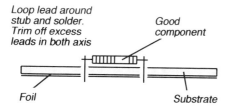

NOTES:
(1) Cut leads as close to component body as practical
(2) Straighten remaining wire stubs taking care to avoid damaging the track
(3) Cut leads of new component to lengths required and loop each wire once around stubs
(4) Solder new component leads to wire stubs using correct heat and low melting point (LMP) solder
(5) Trim excess leads to avoid shorts and protruding wires
(6) Check solder joints and track for damage

Fig. 7.1 Replacing defective component – preferred method.

Method 2

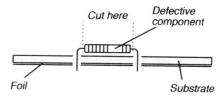

NOTES:
(1) Remove component by cutting leads as indicated
(2) Unsolder wire stubs using suction or capillary tool to collect surplus solder
(3) Clean area if necessary
(4) Remove old wire stubs carefully to avoid track damage
(5) Insert new component ensuring that the body of the component is spaced correctly from the board
(6) Cut new component leads to correct lengths and solder in position

Fig. 7.2 Replacing defective component – alternative method.

2. MULTIPIN/LEAD COMPONENTS (INCLUDING INTEGRATED CIRCUITS)

Multipin/lead components are difficult to remove without damaging the PCB. There is a high risk of track damage or lift when the heat applied during de-soldering is concentrated in small areas. Wherever possible plug-in components should be used.

B – Repair – broken, damaged or missing track

NOTES:
(1) Choice of repair method depends upon the size of gap between the points to be bridged. Scratches are conveniently repaired using the method illustrated in Fig. 7.3
(2) Remove lacquer, solder resist, etc., with a sharp instrument (e.g. scalpel), fine abrasive paper or glass fibre brush

(3) Table 7.1 below indicates the method to be used according to the gap in the track

Table 7.1

Method	Gap size (mm)			
	0–10	10–20	20–40	40 +
Fig. 7.3*	✔	X	X	X
Fig. 7.4*	✔	✔	X	X
Fig. 7.5*	✔	✔	✔	X
Fig. 7.6*	✔	✔	✔	✔
Fig. 7.7a	✔	✔	✔	X
Fig. 7.7b	X	X	X	✔
Fig. 7.7c	✔	✔	✔	✔

KEY:
✔ suitable repair method
X unsuitable repair method
* Not suitable for the repair of double-sided boards or boards having narrow (0.25–0.5 mm) tracks and spacing

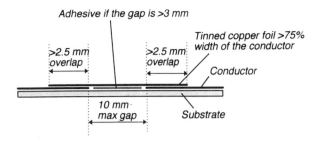

Fig. 7.3 Repair for scratches or gaps in track up to 10 mm wide.

C – Repair – scratches and gaps in track up to 10 mm

NOTES:

■ A piece of pretinned copper foil between 0.035 mm and 0.070 mm thick (This is suitable for boards with 30 g per 900 cm^2 (0.035 mm) copper. For 60 g or 90 g copper proportionately thicker foil should be used.) should be soldered to the track with a minimum overlap of 2.5 mm on each side as shown in Fig. 7.3

■ Where the gap to be bridged is greater than 5 mm the foil should be bonded to the board with suitable epoxy adhesive (see Section 6) (the film adhesives method is convenient for narrow gaps)

■ Foil is preferred for repair of scratches but where track widths are less than 0.5 mm the foil strips become difficult to handle and the repair may have to be made with suitable thickness tinned-copper wire

D – Repair – gaps in track up to 20 mm

NOTES:

■ Bridge gaps with bare tinned-copper wire soldered to track on each side of gap
■ Solder wire along whole width of each overlap
■ Solder fillet to be bright and smooth in appearance and have a low contact angle between track and wire
■ For tracks wider than 1.5 mm use 0.8 mm diameter wire

Fig. 7.4 Repair for gaps in track up to 20 mm.

E – Repair – gaps in track up to 40 mm

NOTES:
This type of repair is similar to the previous figure. However in this case, 0.5 mm diameter insulated copper wire is used or 0.5 mm copper wire with a suitable sleeve fitted.

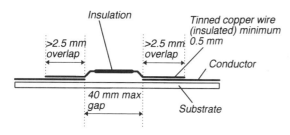

Fig. 7.5 Repair for gaps in track up to 40 mm.

F – Repair – any gap size in track

Single-sided boards only

NOTES:

- This method is essentially similar to Fig. 7.4 except that insulated tinned-copper wire, or wire fitted with a suitable sleeve, is run on the component side of the board via holes drilled in substrate adjacent to track
- If there is any possibility of the insulated portion of wire lifting it should be secured to the board with approved adhesive at approximately 25 mm spacing
- Bare tinned-copper wire suitably sleeved may be used in place of insulated wire

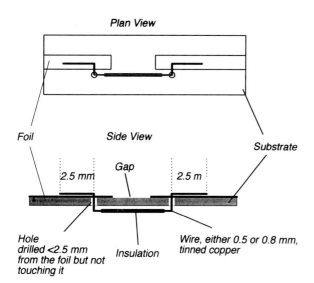

Fig. 7.6 Repair for any gap size – single-sided boards only.

G – Repair – alternative track repair

Single or double-sided boards

NOTES:

- Bridge gap with insulated tinned-copper wire as in (c) or loop and solder to a component lead to give electrical continuity and mechanical strength as in (a) and (b)
- Wire may be run on either side of the board except with gaps greater than 40 mm where the wire should run on the component side as in (b) and (c)
- Irrespective of the method used the existing joint must be unsoldered and molten solder removed before resoldering
- If there is any possibility of the insulated portion of the wire lifting, secure to the board with an approved adhesive at approximately 25 mm spacing
- This method should only be used when the existing component leads protrude by at least twice the diameter of the bridging wire

(a)   Gaps less than 40 mm

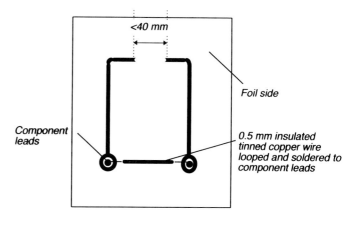

(b)   Gaps greater than 40 mm (see (5) above)

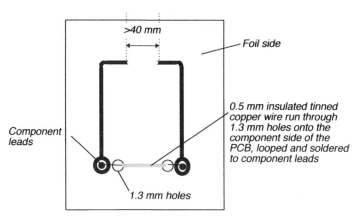

Fig. 7.7 Alternative repairs – single or double sided boards.

H – Repair – track lift

Distorted track lift

NOTES:

- Remove lifted portion of track by cutting at points at least 5 mm on either side of lift
- Remove lifted portion of track
- Restore track continuity using one of the methods illustrated in Figs. 7.4–7.7

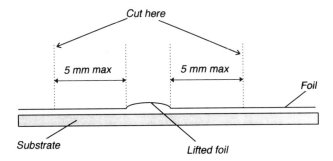

Fig. 7.8 Repair of distorted track lift.

Non-distorted track lift

NOTE: Restore bond between track and laminate using an approved adhesive (see Section 6).

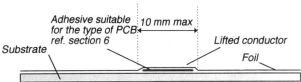

Fig. 7.9 Repair of nondistorted track lift.

J – Repair – damage to plated edge, contacts, etc.

Repairs to plated edge-contacts

Repairs to any part of a PCB where it enters an edge connector are not permitted.

Mechanical repair

Repairs to any part of a PCB which has come chipped, cracked, fractured or defaced are not permitted.

K – Multilayer printed board repair

With the expanding electronics industry and the further integration of designs comes the reduction in size. Tracks and gaps become smaller and increase the possibility for breaks to occur within the circuit. The cost of the basic board and the labour which has gone into its production does mean that repair is a viable concern. It becomes even more viable after it has been populated with components. In most cases, cost is not the only consideration. At this stage, delay can be the most important factor.

Prior to the modification or repair of multilayer boards a number of areas need to be considered:

- Will the method of modification have an effect on the electrical performance of the equipment?
- How will the fault location be identified (a common method is provided by X-ray examination)?

■ Alternatively, if the fault occurred at a manufacturing stage, the production artwork may be
used as a reference.

The use of parallel gap welding is not new to the
circuit industry. It has been used in some companies
for many years. Today many systems are available
which can repair basic and populated boards, both
multilayer and double-sided circuits, without the
repair being detectable.

Due to the need to rework multilayer tracking
below the surface of the board, restrictions are
imposed by the available space, hence the advantages
of gap welding which take advantage of the minimal
access to the conductors.

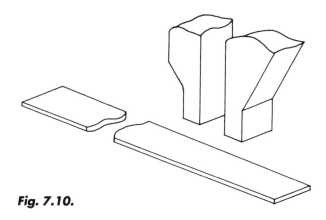

Fig. 7.10.

1. PROCESS

The process of parallel gap welding requires a combination of heat, pressure and time to be brought
together to weld materials. The process is a fusion/
braze weld where the interface material is heated by
resistance to produce sufficient heat to melt a third
metal such as gold which alloys to both materials to
form a permanent bond.

The resistance weld/braze falls between both
extremes and the three variables of heat, time and
temperature are all needed to produce the required
result. The electrode pressure is used to force the
metals together and crack the oxide layer. Heat is
generated by the resistance of the weld materials to
the flow of current. Depending on the time, this will
melt the coating to form an alloyed joint. When the
flow of current stops, the pressure is still applied
during the rapid cooling phase which is usually in
only milliseconds.

Parallel gap welding is a form of resistance welding
where both electrodes are placed on top of the
materials to be joined. Parallel gap welding is a series
weld in which the spacing between the electrodes is
usually less than 0.635 mm (see Fig. 7.10). In this way
a single weld is formed between the electrodes. The
bulk of the electrodes and the repair materials act as a
heat sink and limit fusion to approximately 75% of
the electrode gap.

2. MATERIALS

Copper is the base material for circuitry on basic
boards and it is this material which is to be repaired.
The most commonly used material is gold-plated
Kovar, usually 0.0762–0.127 mm thick. The gold
plating stabilises the surface condition and reduces
contact resistance. During the joining operation it is
the gold which acts as the alloying material and helps
to create the permanent repair.

One of the problems with the use of Kovar is the
expense. Also it is only available in a limited number
of sizes to suit the track patterns. Depending on the
thickness of the gold plate, the base material can also
cause poor soldering. An alternative is the use of
copper foils with a tin/lead coating. This is very easy
to form into the required size to suit all circuit patterns and it will provide an adequate repair. This
form of repair medium can cause a problem in its
consistency, due to the range of machine settings
which will produce a solder joint, but not form a
permanent bond to stand up to flow-soldering
operations.

A combination of copper foils can be chemically
milled to form any shape and size, with and without
pads. A gold flash of 0.00025 mm will produce a
satisfactory repair which will stand any subsequent
production operations.

The copper foils allow the circuit house to produce
its own foils to specific requirements, very economically. The size and shape can be made to accommodate all the requirements of modern circuit
designs.

3. SPECIFICATION

The use of this method of repair is now recognised by
both the British Standards and military specifications

in the USA. The method of interconnection has been used and approved for assembly of flat packs on printed boards, this being one of its original uses.

The current issue of BS 6221, Part 21, which relates to repair methods for printed boards, covers methods of repair by parallel gap welding in great detail. Full evaluations of the use of the repair method by customer inspectorates has continued to confirm its quality and reliability.

With regard to quality, this form of rectification can be invisible when produced at board level and is the only method of repair which can be effectively used on track sizes below 0.25 mm. This is because of difficulties caused by the size and handling of conventional repair materials.

4. REPAIR METHOD

The precise repair sequence will vary with the type of boards being repaired. With multilayer boards where inner tracks need to be remade, the board will require routing through the surface epoxy down to the broken conductor. With conventional boards the solder resist will need to be removed. Removal of the substrate material to the level of site which may need repair or modification is best achieved using a routing machine and routing bits. This allows a milling operation to be conducted with accurate control of both x and y movement and the depth routing.

After the initial steps of board preparation a normal sequence of operations for gap welding can be as follows:

(1)   Check the electrodes by cleaning and ensuring they are level. These two actions can be done using a piece of silicon carbide paper of approximately 800 grits. Placing the paper under the electrodes with the heads down allows the paper to be drawn forward against the electrode faces. This cleans them and checks that they are level. This is indicated by two parallel lines drawn on the paper.

(2)   For welded repairs using gold-plated copper foils, the area to be repaired must be clear of solder or tin/lead coating. If this is left in place a solder joint may be formed rather than the more durable welded repair. The tin/lead coating can be removed with a combination of knife blade and glass fibre brush. It will be necessary to abrade the coating and some of the copper, therefore care should be taken to avoid removing too much copper.

(3)   The correct size repair foil can then be selected with the required weld voltage, time and electrode gap. A representative test board should be used to check the settings prior to conducting the repair operation. The board can then also be used by QA/QC staff as a sample for whatever evaluation may be required to confirm the satisfactory repair. In the case of a gap weld a dark line should be visible across the position of electrode contact. This visible sign is an indication of a satisfactory weld.

(4)   To form the weld repair, the repair strip is positioned on the track across the defect, ensuring that the foil is aligned with the base conductor. The electrodes are then lowered holding the foil in position, and when the force is applied the voltage is triggered to form the weld. This operation is repeated on the other side of the repair area (see Figs. 7.11 and 7.12).

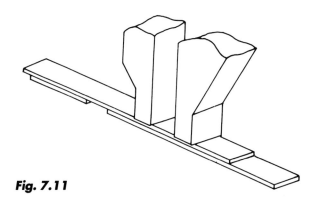

Fig. 7.11

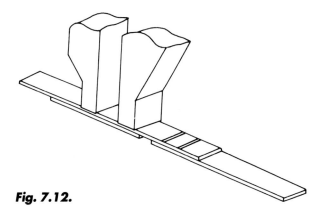

Fig. 7.12.

(5)   On the completion of the repair, the foil may be tined using a soldering iron. This will make the completed repair virtually invisible. In the case of inner layers the tinning operation is not conducted.

(6)   The final stage of the repair operation is to replace the material which was removed to gain access to the repair site. Generally a mixture of

epoxy resin is used to cover the tracking and fill the void which remains. This operation is best achieved when the board temperature has been raised to allow full penetration to occur. On completion any resin which is above the surface after cure can again be removed using the routing machine.

One of the first considerations about the repair is the possibility of lifted pads. The small size of tracks and pads on most surface mount boards requires the minimum of stress imposed on pad areas when removing components. This should be considered when assessing equipment. The force applied during conventional repairs will be supported by the plated-through holes. With surface mount technology (SMT) packages the only strength which is present is the bond between the conductor and the board, which is at its weakest during the desoldering operation. A further consideration is the possibility of adhesive which may have been used to hold the component in position during assembly. Failure to consider this may result in lifted resist or tracks which will require further work to rectify.

Section L is a suggested step-by-step guide to the reworking of surface mounted devices. The steps may require some modifications on specific boards.

These methods describe the removal of chip components, normally referred to as surface mounted devices. A section is devoted to each type of device which may have been soldered to a printed board. After removal, the reuse of components depends on their type and any visual damage which may be apparent. Components which have metalised terminations such as resistors and capacitors will, after component removal, suffer some degree of loss of metalisation. These devices should not be reused. Components which are fitted with leads, such as transistors and integrated circuits may be reused, provided that the component is not damaged during removal. Resoldering of leaded devices may be difficult if the leads are poorly seated because of movement which occurred during removal. The device should be discarded if the leads cannot be levelled to fit the board circuit as damage may have occurred to the tracking.

It should be noted that some components may be bonded to the board by adhesive. In the case of resistors and capacitors, heat applied to the board to reflow the joints will break the bond. Leaded devices require complete lead removal prior to breaking the adhesive bond.

L – Surface mount repairs

1. INTRODUCTION

SMT is enabling the electronics industry to make significant steps in the manufacture of board assemblies. The assembly is becoming miniaturised to such an extent that repair or rework is difficult, but not impossible. Selection of the right equipment, backed by detailed training, can make the process easy to perform.

Consideration for repairs must be yet another of the close cooperations between design and production departments to allow equipment to be used in an effective manner. The way boards and assemblies are designed can have a major effect on the ease of reworking. Repair tools, depending on the type, may need access around the components. Liaison between the design office and production staff can ensure that the correct equipment is specified.

Repair of surface mounted components can be easy, if the equipment and the operator training have been fully considered. As in any operation, training is the key to producing a satisfactory product.

When considering rework equipment, it is fair to say that one tool will not do all the jobs: it may be necessary to buy more than one piece of equipment. The choice and set-up of repair stations will depend on many factors, and each should be considered prior to making the final choice. Types of components, types of substrate, production techniques, reflow/flow-solder, design of assembly.

Present day surface mount components use lead counts which can range from 2 to over 100 with lead spacing between 0.10 and 0.125 mm. Future development will see the lead count increase with lead spacing decreasing. Both areas make it difficult for inspection and rework.

2. QUALITY/INSPECTION

The quality of welded repairs is visible directly by visual examination for a weld trace. This shows as a dark line across the repair foil which indicates that the repair has been completed.

The quality of this type of repair can be confirmed on sample welds by microscopic examination of metallurgical cross-sections. A simple test for satisfactory welds on test boards is to try to remove the foil by peeling it back after repair. If a satisfactory repair has been produced either the base/repair foil will break or both will be removed from the board. A satisfactory weld will exhibit joint strengths which exceed the printed circuit pattern adhesion.

Because of the heat generated by the parallel welding operation, the epoxy bond between the copper foil and base fibre glass will loosen. This can be minimised by varying the weld duration settings. Minimal time will limit heat, thus preventing epoxy spread.

Excessive weld duration will be shown by an excess of epoxy adhesive visible at the weld point. This should be minimised.

The following points should be considered during the quality control of this method of board repair:

- Operators should be properly trained in the use of the equipment and methods of repair
- The repair foil should be the same size as the track under repair and both should be aligned
- The basic board should show no signs of voids around the weld. A limited sign of epoxy melting around the fillet is acceptable
- The board track or foil should not be reduced in width at the repair area
- Before tinning the repair area, a weld should be visible as a dark line across the foil.

Parallel gap welding is a very economical and reliable method of board repair. The reduction of standard track sizes below 0.254 mm makes convention repair methods impractical. In the case of multilayer boards, other methods of repair and modification are not practical.

3. REMOVAL OF INTEGRATED CIRCUITS, TRANSISTORS, DIODES

These devices are generally encapsulated in plastic with flat leads.

Method 1

(1) Hold the board firmly to allow free access ensuring no damage
(2) Remove protective conformal coating if necessary
(3) Using a miniature soldering iron, remelt the solder from around the first component lead and lift the lead from the board using a special probe. Care should be taken to apply the iron

for just enough time to free the lead
(4) Repeat (3) until all leads are free of the board, and remove device. Great care must be taken to avoid lifting any conductors
(5) Using a desoldering tool, remove any excess solder from the lands which may prevent the component from being reseated
(6) Inspect the lands for signs of lifting. If the copper is found to be lifting, the board must be repaired prior to fitting
(7) Tin the replacement component leads if a tin lead finish has not been used, position component and resolder, ensuring that each terminal is heated for only the minimum time required to effect a successful joint. Flux should be applied to the pad and lead areas prior to reflow. Use a flat probe to hold the component in position until the solder has reflowed and solidified.

Removal of resistors and capacitors

These devices are generally a ceramic material with metalised termination. When soldering or replacing ceramic components with metalised terminations an LMP solder must be used. The solder should also include a silver content of approximately 2%.

(1) Hold the board firmly to allow free access, ensuring no damage
(2) Remove protective coating as necessary
(3) Using a hot gas desoldering system, reflow the solder joints. When the solder is molten, simultaneously remove the component with a pair of tweezers
(4) Using a desoldering tool, remove any excess solder from the land
(5) Inspect the lands for signs of lifting. If the copper is found to be lifted, the board must be rejected and a concession obtained to repair the defect
(6) To replace the component, tin the end termination if it is not tin/lead finished. Resolder the new device into position using the hot gas system. Ensure that each terminal is heated for only the minimum time required to effect a satisfactory joint. Flux should be applied to the land area prior to reflow
(7) Remove flux with a solvent cleaner.

Method 2

Ceramic devices such as resistors and capacitors may be removed using resistance-heated tweezers. This method allows the solder joints to reflow plus it allows a force to be applied to those devices which are bonded to the board.

It is not recommended that this method be used for replacement.

(1) Hold the board firmly to allow free access, ensuring that no damage can occur
(2) Remove protective coating as necessary
(3) Using resistance-heated tweezers, reflow both solder joints. After both solder fillets are molten, lift the component free of the board surface. It may be necessary to apply a slight pressure where devices are bonded to the board surface
(4) Using a desoldering tool, remove any excess solder from the land area
(5) Inspect the lands for signs of lifting. If the copper is found to be lifted, the board must be rejected and a concession obtained to repair it
(6) Make sure that the leads are placed correctly in all positions
(7) Remove the flux with a solvent cleaner.

Integrated circuits may be removed by the use of a special temperature controlled desoldering head which reflows all leads at the same time. In operation all leads are reflowed and suction force by the desoldering station lifts the device from the board. This method may not be satisfactory where the component has been bonded to the board; in this case Method 1 should be used.

Method 3

This method is more suitable for transistors and diodes which only contain three leads.

(1) Hold the board firmly to allow free access, ensuring no damage
(2) Remove protective coating as necessary
(3) Using a hot gas reflow system, reflow all leads simultaneously. When the solder joints are reflowed, remove the component with a pair of tweezers
(4) Using a desoldering tool, remove any excess solder from the land
(5) Inspect the lands for signs of lifting. If the copper is found to be lifted, the board must be rejected, and a concession obtained to repair the defect
(6) To replace the component, tin the leads, if not tin/lead finish, then resolder the new device into position using either the hot gas system or a miniature soldering iron. Ensure that each

terminal is heated for only the minimum time required to effect a satisfactory joint. Flux should be applied to the land area prior to reflow
(7) Remove the flux with solvent cleaner.

4. DESOLDERING EXAMPLES

Example of desoldering operation using conductive tweezers

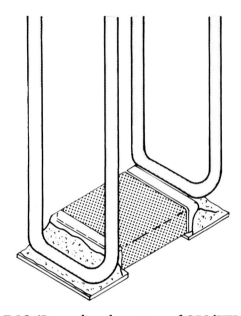

Fig. 7.13 (Reproduced courtesy of GEC/EITB).

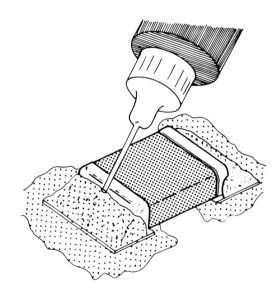

Fig. 7.14. (Reproduced courtesy of GEC/EITB).

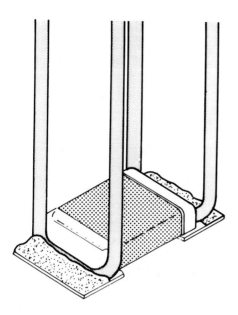

Fig. 7.15. (Reproduced courtesy of GEC/EITB).

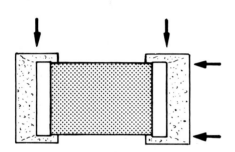

Fig. 7.16. (Reproduced courtesy of GEC/EITB).

Example of the desoldering operation using soldering iron.

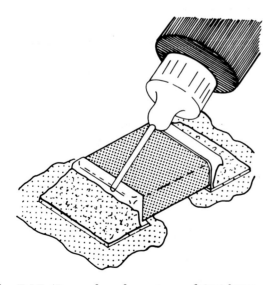

Fig. 7.17. (Reproduced courtesy of GEC/EITB.

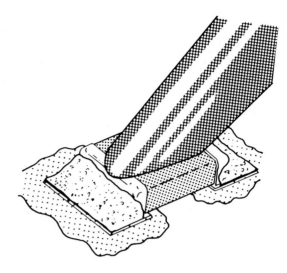

Fig.7.18. (Reproduced courtesy of GEC/EITB).

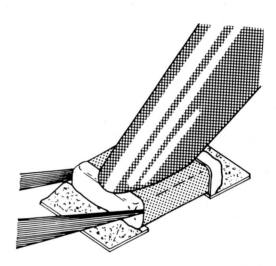

Fig. 7.19. (Reproduced courtesy of GEC/EITB).

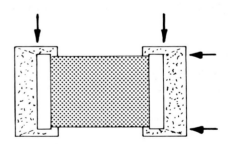

Fig. 7.20. (Reproduced courtesy of GEC/EITB).

Example of the desoldering operation using the hot gas pencil.

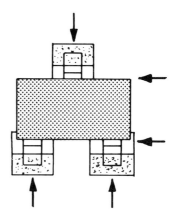

Fig. 7.24. (Reproduced courtesy of GEC/EITB).

Example of the desoldering operation for multileaded components.

Fig. 7.21. (Reproduced courtesy of GEC/EITB).

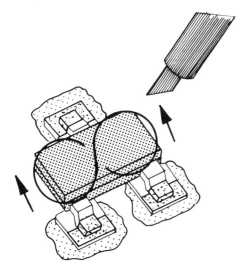

Fig. 7.22. (Reproduced courtesy of GEC/EITB).

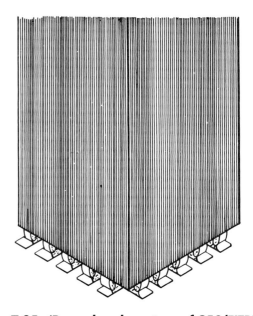

Fig. 7.25. (Reproduced courtesy of GEC/EITB).

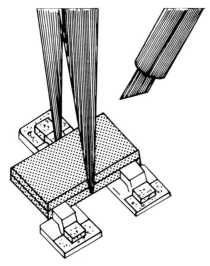

Fig. 7.23. (Reproduced courtesy of GEC/EITB).

Fig. 7.26. *(Reproduced courtesy of GEC/EITB).*

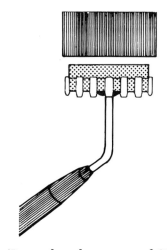

Fig. 7.27. *(Reproduced courtesy of GEC/EITB).*

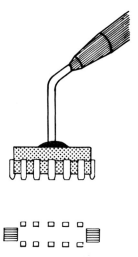

Fig. 7.28. *(Reproduced courtesy of GEC/EITB).*

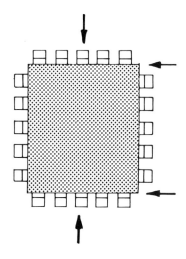

Fig. 7.29. *(Reproduced courtesy of GEC/EITB).*

SECTION 8: DEFECTS CLASSIFICATIONS

A – Defects classifications

NOTES:

- Defects may be either Class A, B or C defective but not any two together, in this case the most severe of the two classifications should be used
- Two or more simultaneous Class C defects shall be a Class B defect
- Two or more simultaneous Class B defects shall be a Class A defect

CLASS 'A' DEFECT

A defect more serious than one categorised as Class B and which renders a component, subassembly or assembly unsuitable for use and including those manufacturing defects which if discovered during manufacture, installation or when first put into service (e.g. unsoldered joint) would entail attention or corrective action.

CLASS 'B' DEFECT

A defeat likely to cause failure or additional maintenance and indicative of an unacceptable standard of workmanship or production.

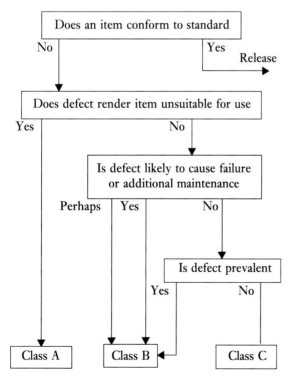

Fig. 8.1 Algorithm.

CLASS 'C' DEFECT

A defect other than Class A or B where the prevalence of a Class C defect indicates an unacceptable standard of workmanship or production.

B – Surface mount soldering fault chart

1. INTRODUCTION

The use of new technology in many production areas provides a number of problems for engineering staff. SMT has provided a number of new faults which may become apparent on SMT assemblies.

Table 8.1 covers some of the problems which may be encountered during production and provides possible solutions.

Table 8.1

Condition	Cause	Remedy
Solder balls	• Poor quality solder paste • Incorrect reflow profile • No prebaking before vapour phase • Poor paste storage condition	• Contact supplier • New profile • Bake paste • Contact supplier
Solder wicking	• Uneven heating of assembly • Poor solderability • Limited solder paste	• Prebake assembly • Contact supplier • Check screen print
Tomb-stoning	• Poor solderability • Poor pad design • Uneven heating profile • Resist thickness	• Check storage conditions • Correct design rules • Check preheat steps • Specify thin resists
Open joints	• Poor solderability • Limited solder paste • Flow-solder shadowing	• Check storage condition, test solderability • Check screen mesh area, printer set-up • Check part orientation, correct pad design, poor preheat profile
Leaching	• No barrier layer • Temperature too high • Soldering operating too long	• Specify correct termination • Reduce temperature • Reduce time
Voids	• No solder paste bake • Adhesive contamination • Poor solderability • Too much flux	• Bake solder paste • Check application • Test solderability • Test flux application

SECTION 9: QUALITY MANAGEMENT SYSTEMS

A – Introduction

Much has been said and printed about quality management system (QMSs), particularly when the accreditation to ISO 9000 is involved. Within this section the benefits, and equally importantly the drawbacks, of having a documented QMS are investigated. Each company must look at its own reasons for developing a QMS, and decide how far it intends to pursue this form of management control.

A QMS is a tool to be used by a company to ensure that the quality of the goods or product they delivery is consistent. The QMS provides a method for ensuring that the shortcomings of the system and the product are brought to the attention of the right people, and that they have the tools to resolve the problem, permanently. Used properly it will assist managers to improve the quality of the product and maintain the quality at the desired level. That said, it needs commitment at all levels, from the managing director (MD) to the people on the shop floor, to make it work.

B – Selecting a Standard

References to ISO 9000 in this section, should be taken to mean ISO 9001 unless otherwise stated.

Benefits

A QMS will provide the user with a documented management system. From the information it provides and the processes it drives, the handling of orders or tenders and the service contracts to maintain it can be controlled from the point they arrive, to delivery to the customer.

The QMS enables its uses to demonstrate that they have independently verified internal controls and that the quality of the product is consistent.

Implemented properly, it will allow managers to know when things are working well, and when they are not. When they are not, the open nature of a QMS should permit the people best suited to suggest improvements to do so. A correctly implemented

QMS is one of the key stages to a process of continuous improvement.

Drawbacks

The key drawback is cost. Whichever way one chooses to look at it, a QMS will cost money, the cost falls into three key areas:

- Cost of accreditation, and maintenance of that accreditation
- Additional internal administration
- Changes in working practice

If a company chooses to go down the route of the QMS it will need to decide which of the standards should be adopted.

ISO 9001: This standard should be adopted if the company is involved in designing products (not making them to someone else's drawings).

ISO 9002: This standard should be adopted if the company is manufacturing products to other people's designs and has no in-house design function.

If the company is developing software, ISO 9000-3 should be consulted.

C – Quality manual

In small companies, this manual would normally contain a list of all of the procedures. In larger companies it would normally contain the top-level procedures. In both cases it shall contain the quality policy and a matrix showing which procedures comply with which clause in ISO 9000.

D – Management responsibility

Quality policy

The MD or equivalent should draw up a quality policy for the company. The quality policy should be short, concise and relevant to the business. It defines the company's objectives for, and commitment to, quality. This policy should be published so that all employees are aware of it.

Organisation

There should be a simple organisation chart showing the structure of the company. A typical chart would show each layer of management down to department level and the management representative separately, with a dotted line linking him/her to the MD, regardless of his/her other duties.

The formal responsibilities within the QMS should be defined in the 'responsibilities' section of each procedure, the quality manual, and within each person's job description which should be reviewed annually.

Management review

The effective implementation of the QMS should be reviewed during the formal 'management review'. A management review should consist of a systematic review of the information provided by the processes within the QMS, and provide actions for any shortcomings. Details of the information required should be contained in the individual procedures. The review should be carried out in one meeting, or separately and the results should be summarised and presented at the management review meeting. Management reviews shall involve the senior managers, be minuted and be carried out at least quarterly.

E – Quality system

General

In implementing a QMS the company is automatically conforming to this section.

Quality system procedures

The simplest way of carrying this out is to have procedures that cover each of the clauses of the standard that has been adopted. A quality manual will also need to be drawn up. Most small companies include their health and safety procedures in the system. Further components of this section will show what should be included in each procedure.

Quality planning

This procedure should ensure that the company's preferred methodology in terms of planning is defined, and if there are any templates or standards to be adopted, such as pro-forma project plans or quality plans, these templates shall be controlled as if they were procedures.

Section 4.2.3 in ISO 9001 should be read for information regarding planning. However it is an established fact that careful planning pays off and when constructing project or quality plans, the following points should be borne in mind:

- Have a concise working document
- Keep to facts that can be verified
- Avoid unnecessary controls
- Keep the correct records
- Avoid unnecessary paperwork

F – Contract review

Review

This procedure should serve to separate 'professional' organisations from those that say 'yes we can do it' and then sort out the problems after they have taken on the work.

The procedure should ensure that the contracts (or tenders) are read by the most appropriate person, who should sign or initial it to say that the company:

- Has the resources to carry out the work
- Has documented and confirmed verbal instructions
- Has ensured that the details of the order are as clear as is reasonable to expect

For small companies, the acceptance of verbal orders is permissible for small jobs; however, in these cases, the order should be written down and repeated back to the person giving the order, to avoid mistakes.

Amendment to a contract

This procedure should show how any amendment to

a contract takes place and how it is reviewed for the same reasons shown under 'Review' and by the same people or departments who originally reviewed the contract.

Records

The procedure shall show how records of such reviews shall be kept.

G – Design control

The procedure for design control should be broken up into its constituent parts as shown below. It is perfectly reasonable to use this procedure to control how design plans are drawn up and controlled. Indeed this will permit a more flexible and innovative approach. As this is the least bureaucratic approach, it is the example shown below.

Design and development planning

There should be a plan that takes the product through its life-cycle. A typical life-cycle for simple bespoke designs is:

- Review tender
- Initial design proposal
- Conformation with customer
- Final design proposal (statement of requirements)
- Design work
- Design reviews
- Final design review
- Freeze design

Individual life-cycles will vary depending on the complexity and source of the work.

Organisational and technical interfaces

The plan should contain the organisation chart showing:

- Project manager
- Groups involved within the company
- Interfaces between those groups
- Interfaces between the company and any external agencies or customers

Design input

The plan shall show any contractual, statutory, regulatory or other design requirement, that should be adhered to, such as:

- IEE regulations
- Company standards
- Design specifications from the customer
- Environmental specifications
- Safety standards
- C.E. marking

Design output

The final design should be reviewed before it is output or published, see 'Design review' below.

Design review

As part of the plan, the design authority shall plan design reviews either by date, or when certain criteria are met, such as completion of tasks. These reviews shall be minuted and they shall address the following topics as appropriate. Does the design:

- Meet the original requirements?
- Meet output acceptance criteria?
- Meet safety and functionality requirements?
- Meet handling, operational and environmental requirements?
- Meet maintenance and disposal requirements?

The minutes shall consist of actions required to ensure that the above criteria are met.

Design verification

As part of the process of design, it will be necessary to carry out checks or 'verifications' to ensure that the design will meet its original requirements. These checks should include:

- Alternative calculations
- Comparative analysis with proven designs
- Modelling
- Tests and demonstrations
- Reviews (see above)

Design validation

Design validation is different from verification in that

it is usually performed on the final product or prototype. Design validation should take the form of a test plan, showing all the tests required to ensure that the final product meets its original requirements. A good example of this is the development of aircraft, where design verification is the way to ensure that the aircraft will fly if it is correctly assembled. Design validation is where the test pilot finds out if the aircraft really does fly.

Design changes

At all stages of the design, it may be necessary to change either the specification, or the design itself. This part of the procedure documents the way changes are reviewed by all affected parties, and how the reviews are authorised and implemented. The standard method for carrying out this process is the design change request form (DCR). It is normal to copy the DCR on coloured paper when it is approved to avoid mistakes. The DCR shall show the following information:

- Originator
- All departments affected
- Proposed change
- Actual change if different
- Any attached drawings or documents
- Approval by those required to approve it

The DCR would normally follow the path and process shown below:

(1) Originator completes requested change
(2) Design authority approves initial change and indicates which department shall approve the DCR
(3) DCR is circulated to departments and also to any mandatory departments such as contract department
(4) Finally the DCR is returned to the design authority for approval and implementation
(5) The DCR is then copied on coloured paper and returned to affected departments

A record of all DCRs must be kept.

H – Document and data control

Document and data approval and issue

This procedure defines how documents and data are controlled within the company. Its main purpose is to ensure that any documents used with the company are at a known issue or version level and that obsolete documents are removed. It should apply to:

- Procedures
- Drawings
- Work instructions
- Design specifications
- Plans

Specifically, the procedure should refer to a master index of all procedures and work instructions within the company. This index is the key to the document control system and should hold the current version number of all procedures and work instructions.

It is normal for the QMS to control templates and standards within the company. One of these standards should be a template for minutes in meetings. There is a suggested template at the end of this section.

Document and data changes

This part of the procedure shall show how documents are correctly reviewed and authorised. In general the same people or departments should review the same procedures or documents where this is practical.

Where practical the changes should be shown in the new documents. Word processing makes this a simple task.

It is normal for this procedure to carry a pro-forma procedure as a template. In this manual it is shown at the end of this section.

To ensure that procedures are received and implemented when new revisions are published, it is normal within a QMS to send all procedures out under the cover of a transmittal note. The transmittal note is a receipt that is signed and returned to the department responsible for distributing procedures, to confirm receipt and destruction of old procedures. A log of all transmittals sent and returned, shall be maintained. All discrepancies must be followed up.

J – Purchasing

Evaluation of subcontractors

This procedure defines the criteria for selecting subcontractors. These criteria include:

- Their ability to meet the requirements
- Any required accreditation or financial constraints
- Any internal requirements such as evaluation audits

This procedure also defines the process for keeping an accurate record of acceptable subcontractors, together with the scope of work for which they are acceptable. A list of contractors who have been removed from that list should be kept with reasons why they were removed. The reasons should be clear and objective, so that good subcontractors are not prevented from retendering if they were removed solely because of time. If this information is kept on a computer, the Data Protection Act may apply.

Purchasing data

This part of the procedure should refer to a proforma purchasing form which should have space for the following information (where applicable):

- Exactly what is required including grade or size
- Identification of any drawings, instructions or specifications and issues
- Inspection instructions, to which the goods will be subject, particularly if the inspection is to be carried out at the supplier's premises
- Any special processes, and the standard to which they should be carried, e.g. galvanising to the appropriate British Standard or company standard.

The procedure shall require this data to be entered on the form when necessary.

Verification of purchased product

See 'Purchasing data' above. This part of the procedure defines the method for getting product released after inspection.

Customer verification of subcontracted product

Where required by the contract, this procedure should define the way the customer verifies the supplier's product. This process is usually defined in the appropriate project, contract, or design plan, however, the process should be required in this procedure.

K – Control of customer supplied product

Customer supplied product is product supplied free of charge by the customer, or in some cases where it is purchased at the specific request of the customer. A typical example is the customer supplying 1000 steel blocks to be milled into a specific shape, or requesting the purchase of 1000 steel blocks for the same reason and agreeing to pay for the blocks as a separate part of the contract.

This procedure defines the process for ensuring that such product is:

- Identifiable as owned by a particular customer
- Stored appropriately
- Maintained if required
- Returned or reported to the customer if it is lost, damaged, or found unsuitable

Records of the above must be maintained.

L – Product identification & traceability

Product identification and traceability is dependant on the type of product being produced, and each company should decide together with its customer what those requirements are.

In a complex product i.e. motor cars, or where safety is involved, it may be required to trace individual items back to their batches, as supplied by their respective suppliers. If the company produces boxes for toys, it might only be necessary to record incoming card and ink stocks so that faulty batches could be returned to the supplier in their entirety.

M – Process control

Process control is normally carried out under the control of individual work instructions and procedures. The pro-forma procedure referred to earlier in this section and shown at the end should be used as a pro-forma work instruction, although it could be simplified.

Repetitive tasks, specifically those required to meet standards or use specific tools, torques or other defined workmanship standards should have the process controlled by work instructions or drawing.

As a general rule, if the absence of a procedure might result in a reduction in the quality of the product, then a procedure or work instruction *must* be put in place. In some cases, company standards or the use of manuals such as this one can reduce the need for work instructions.

Maintenance of equipment that affects quality shall be included, although the procedure may simply control maintenance contracts if this is the case. It is also permissible to have some tasks defined in the project or quality plan if one exists. This is particularly important for project related work.

N – Inspection and testing

Receiving inspection and testing

This procedure defines the process of inspection and testing goods received into the company. It should allow for goods to be held in quarantine prior to inspection and test, and for return of goods which fail to their supplier.

If incoming goods are required so urgently that inspection and testing cannot be carried out, the

procedure shall show what retroactive action shall be carried out to limit the risk. This may include special tracking of that particular batch.

In-process inspection and testing

This part of the procedure defines the process for inspection and testing of product during production. If the product fails it should define the actions to be taken, and the persons who have authority to stop or start production. This process maybe defined in specific quality plans or work instructions.

Final inspection and testing

This part of the procedure defines the process for carrying out final inspection and testing and should refer to the relevant design, or quality plan if appropriate.

Inspection and test records

This part of the procedure defines the records required to be kept after inspection. If there are legal or contractual requirements to be adhered to, the procedure should refer to these.

It is normal to issue inspectors with individually numbered stamps. These stamps become the signatures of the inspectors, and as such should be carefully controlled. The use of numbered stamps ensures that tracing and confirming inspection status is fast and simple.

P – Control of inspection measuring & test equipment

Any inspection, measuring, or test equipment used to test or process product, shall be calibrated.

This procedure ensures that the following criteria are met:

- Equipment can be identified as calibrated
- Equipment that has passed its 'calibrate by' date can be identified
- Equipment is positively recalled for calibration and/or servicing
- Equipment is stored, preserved and handled appropriately
- Equipment is calibrated against a known national or international standard
- Calibration is carried out under the control of work instructions
- Equipment cannot be accidentally or deliberately recalibrated without breaking a seal or dismantling the equipment

This is an onerous process for most small companies and the most effective solution is to subcontract the work to a calibration company which will ensure compliance with ISO 9000.

It is not recommended that calibration of complex test equipment is undertaken unless the people carrying out the work are competent to do so.

It is preferable to use a calibration company that is accredited to the National Association of Measurement and Sampling (NAMAS), although this may prove a little more expensive. Some calibration companies provide a multi-tier service, so that companies can select the most appropriate service.

Q – Inspection & test status

Inspection and test status, in this context refers to the ability to determine the inspection and test status of equipment at any time during production.

For most products on small production lines a tie-on tag with signatures or stamps will suffice. However, for faster production lines or fully automated lines, a more innovative approach may be required. It must be possible to select a product at any stage in production and be certain as to its inspection and test status. This should take into account automated testing and inspection, where the presence of the product at a set position on the line indicates its test status.

R – Control of non-conforming product

Review and disposition of nonconforming product

When any product fails an inspection, or its status is changed because it has been damaged or subjected to unknown conditions (dropped), it becomes nonconforming, and its disposition needs to be determined.

The term 'disposition' frequently confuses and leads to a misunderstanding of what is required. In broad terms, if the product is nonconforming, it should be quarantined, and someone should decide what to do with it (its disposition). The problem may be as simple as being painted the wrong colour or as drastic as being physically damaged. In practice the disposition falls into one of two categories.

Category one

The product does not meet its original requirements but the failure does not affect 'fit, form or function'. In these cases it is normal to apply to the customer for a concession using a concession form from either the supplier or the customer, although a letter will normally suffice.

Category two

The product does not meet its original requirements and the failure affects 'fit, form or function'. In these cases the product should be held, and one of the following carried out:

- Product is reworked
- Concession is granted as above
- Product is regraded
- Product is scrapped

If the product is 'customer supplied product' then refer to 'customer supplied product'. Records of all nonconformities must be kept. The two accepted methods of quarantining product are, to either place it in a marked quarantine area, or mark it with a suitable label or red tape.

In the case of product that has been regraded, it is often possible to continue to use it, although it is known that the results will require a concession from the customer due to some minor defect in finish or similar problem. In these cases the concession should take the form of a permit or concession to manufacture, which should be obtained before any further work is commenced.

S – Corrective & preventive action

Corrective action

This procedure defines how the following issues are recorded, and followed through to conclusion.

- Any customer complaints or positive feedback
- Reports of nonconformities in products
- Reports that the QMS is inaccurate

Preventive action

This part of the procedure defines how preventive actions are followed through, and is usually divided into three areas:

(1) Pre-emptive information of failures that should be treated as corrective actions
(2) The collation of information from audits, corrective actions, nonconformities, customer complaints and positive feedback
(3) Any analysis of data on nonconformities, audits and other feedback for presentation to the management review

T – Handling, storage, packaging, preservation & delivery

This procedure should do exactly what it says. It should define how all product is handled, stored, packed, preserved and delivered.

In most cases, it will refer to the manufacturer's

instruction or require staff to take sensible precautions, follow notices, or if in doubt, consult with the appropriate person.

It may refer to company standards such as those for static precautions or special handling notices.

Delivery instructions shall always carry a handling requirement so that delivery companies are aware of them.

U – Control of quality records

ISO 9000 requires that a number of records are kept to provide evidence that the QMS works and to assist in tracing problems. Some of these records may also be kept as a result of legal requirements.

Apart from the need to be legible and retrievable, there are two established methods of controlling 'quality records'. The first is to maintain a master index, and define only those records as 'quality records'. The second is to define in the quality manual, that records referred to in procedures of work instructions are classified as 'quality records'. The latter is usually the most effective as it allows users to see which records require special attention.

V – Internal quality audits

It is recommended that the staff required to carry out internal audits attend a recognised training course from one of the companies who are part of the Institute of Quality Assurance Registered Assessor Scheme, or BSI.

This procedure defines the method of setting up and carrying out internal audits. Audits should be as short and simple as practical, and the auditors should remember that they are auditing the system, not the personnel.

The results, or an analysis of them, should be submitted to the management review. There must be an annual plan to audit all departments and procedures at least once a year; more important functions may need auditing more frequently to guarantee conformance.

W – Training

Each employee should have a training plan as part of their job description, and this should tie in with a central training plan. It is acceptable to have a single, central, training plan, and this procedure should define the process for identifying training requirements.

Additionally this procedure defines how the company ensures that the personnel who are responsible for the quality of the product are qualified on the basis of the following:

- Education
- Training
- On-job training
- Experience
- Testing

A typical example is where steel fabrication companies use welders who are certified (coded). The supervisors keep a chart of which person is coded to which type of welding, to facilitate allocation of work.

X – Servicing

If the company is responsible for the maintenance of its own or another company's products, this procedure should detail how this is done and the work instructions related to that work.

Y – Statistical techniques

If the company uses statistical techniques, this procedure should detail which techniques are used and how the results are used. Alternatively, it should refer to the fact that this information is detailed in other procedures.

Company name	Procedure number Proc–01	Page of
	Issue No. draft	Effective date

Procedure title **Z. Sample procedure**	Prepared by *A.N. Other* Date	Checked & authorised by *A.N. Other* Date

Index

	Page no.
Purpose	2
Scope	2
Procedure details	2
Responsibilities	2
Definitions	2

Company name	Procedure number Proc-01	Page of
	Issue No. draft	Effective date

Purpose

The purpose section shall contain the purpose of this procedure.

Scope

The scope section contains the areas to which this procedure applies and where appropriate, the areas to which it does not.

Procedure

Within the procedure, describe the process in simple, logical steps.
To define a step as mandatory, use the term *shall*.
To define a step in a process as 'recommended', use the term *should*.

Responsibilities

Detail who is responsible for what within the procedure.

Definitions

A list of definitions used in the procedure should be placed here.

Log number 000001   Date logged _____

AA. Sample design change request form

Requested change

Name _____   Date _____

Reason for request

☐ Department A   Date _____ | These departments must initial to
☐ Department B   Date _____ | confirm acceptance of the design
☐ Department C   Date _____ | change.
☐ Material planning   Date _____ |
☐ Purchasing   Date _____ | Signed _____
☐ Quality assurance   Date _____ | Design authority

Attachments	Cost of implementation
	Parts £
	Hours No. hours
	Recall costs £

Reason for rejection

Design change accepted/rejected* for implementation

* Delete as applicable

Signed _____   Dated _____
Design authority

AB. Sample transmittal note

TN no. 000001

To
A.N. Other
Manufacturing Dept
Ground Floor

From
Document Control
2nd Floor
Tel: 1234

Date:

Please find enclosed the following documents:

Procedure/work instruction/drawing	Version	Copy number

Please replace existing documentation or insert new copies into the relevant folders or in accordance with the attached instructions.

Please complete and return the lower half of this document to confirm that the documents have been received correctly. If there are any discrepancies within the documentation, please return the entire package to the document control department.

I confirm that all the documentation shown on this transmittal note has been received and that superseded documentation has been either destroyed, returned to document control, or clearly marked 'superseded' in red ink.

Signed _____

Date _____/_____/_____

Print name _____

Distribution: _____

_____

_____

TN No 000001

Subject: AC. Sample minutes form	Page:
	of:

Date:		Time:	Record decisions	
Venue:			1	
Chairperson:				
Minutes:			2	
Present:				
1			3	
2				
3			4	
4				
5			5	
6				
7			6	
Next meeting Date:				
Time:	Venue:			

Minutes of meeting

Record of actions:		By:	Record of actions:	By
1			4	
2			5	
3			6	

AD Sample certificate of conformance

Customer details	Company details
Name	Certificate no.
Address	Date
Purchase order no. Contract ref.	

Deliverable description	Qty

Authorised design changes and deviations

Concession no. Production permit no.	Date	Brief description	Batches or items affected

It is certified that all the goods detailed hereon have been inspected, tested and unless otherwise stated, conform in all respects with the requirements of the contract shown above.

Authorised signatory

_____

Distribute to	QA stamp

Minimum distribution –
Customer
File copy

SECTION 10: TERMINOLOGY, ABBREVIATIONS & SYMBOLS

A - General

- Flag terminal
- Ring terminal (flat tag)
- Spade terminal (cable lug)
- Pin terminal (solid)
- Turret terminal (tubular – open or blind)

BA

British Associated

Braiding

Outer earth screen on screened cables or outer conductor of coaxial cable

C of C

Certificate of conformance

Crimp

Wire mechanically clamped to a terminal or connector pin. Many crimped connectors use insulation piercing technique

Def. Stan.

Defence Standard

Crinkle washer

Restrains movement of a nut or screw and preserves surface finish

ESD

Electrostatic discharge

Frettage

Removal of surface by vibrating, rubbing, contact of another component or component lead

GRP

Glass reinforced plastic

Griplet

Device to ensure good metal to metal contact before soldering, especially PCBs

Hank bush

Screw fixing device swaged to a panel or PCB

IQA

Institute of Quality Assurance

Lacing

Plastic cord to maintain shape of cableform

Link

Metal strip or wire joining two parts of a circuit

LMP

Low melting point

Locknut

Thin nut for tightening against a full nut to prevent movement

NAMAS

National Association of Measurement and Sampling

Plain washer

To preserve surface finish beneath nut or screw head

PVC

Polyvinyl chloride

QA

Quality assurance

QC

Quality control

QMS

Quality management system

RF

Radio frequency

Shakeproof washer

Locking washer which bites into surface and nut

SHA

Special handling area

Solder vent

Small hole on one side of a blind turret terminal or thimble to permit air to escape when soldering

Spring washer

To prevent movement of a nut. Often with biting projection

Strapping

Method of binding cableforms to maintain shape. Usually made of nylon with a non-release lock

Swage

Mechanical clamping method of such devices as eyelets, terminals and hank bushes

Thimble

Terminal for terminating a wire end or is attached to an insulating board

Turret lug

Hollow tubular terminal

Whiskering

Strands of wire in a multistrand cable becoming unravelled or thin line of solder causing a short

Wire wrap

An electrical connection whereby the wire is mechanically wrapped around a square section terminating pin

PSI

Pounds per square inch

UTS

Ultimate tensile strength.

B – Wiring

DEFINITIONS

AWG

American wire gauge

Breakout

The point at which a wire or group of wires emerge from a laced portion of a wire harness or cable assembly

BTC

Bare tinned copper wire

Compression crimping

Compression crimping is a method for joining an electrical conductor (wire) to another current carrying member. The compressed juncture is called the 'crimp joint'

Connector – plug

A plug is secured to any wall or panel but is free to move in order to be plugged into a receptacle. A plug may be either male or female in construction

Connector – receptacle

Normally mounted to a surface, a receptacle may be either male or a female construction

Continuity test

An electrical test to determine the presence of a broken connection. The recommended instrument is the ohm meter (Avo) not a buzzer. The buzzer function depends on the battery voltage which is in excess of the permissible tolerance for continuity test; usually the buzzer sounds above one ohm, which is not acceptable. In a continuity test the goal is zero ohms

DIL

Dual in line

Lacing (Tyrap®, etc.)

Lacing is the retaining of wires by grouping in a bundle or designated pattern and securing in various ways. Tyraps® are considered to be a form of lacing

SWG

Standard wire gauge

Twisted pair

A cable composed of two insulated conductors twisted together without a common covering. A twist being one complete turn (360°) of the individual wires

Wire/lead dress

The arrangement of wire/leads in an orderly manner

Wire stripping

The removal of the outside insulation. Three methods are normally used; thermo stripper, mechanical hand stripper and automatic stripping machine

C – PCBs

Laminate

Basic structure material which carries the track on both single-and double-sided boards

Land

Portion of conductive track pattern for connection and/or attachment of components

Plated-through hole

Hole in a double-sided board where the conductive path (track) is taken through the board to the track on the other side

Printed circuit board (PCB)

Board with electrical/mechanical components attached and with all the process of fabrication, soldering, etc., composed

Repair

Treatment of a Class A or B defect such that it can no longer render an item unfit for its intended purpose

Repair (approved)

Repair that has been carried out in a way approved by QA authority in accordance with this standard

Scratch

Break in track which exposes the laminate

Track

Electrical conductive path between components, terminals and contacts

Void

Minute imperfection in conductive track or aperture in a soldered joint

D – Soldering

Bridging

When two or more joints are joined together by a solder fillet or bridge

Cold joint

Joint appears frosty and granulated because of movement during solder solidification. In its worst form, it is a fractured joint

Dry joint

No, poor, or intermittent electrical continuity due to movement after removal of soldering iron while the solder was still liquid or pasty, poor soldering technique, contamination etc.

Good wetting

The solder has feathered out with a bright and smooth joint. It has few or no blow holes

Icicles

The solder joints exhibit small solder spikes rather than a clean smooth joint

Poor wetting

The solder makes a large contact angle with the conductor, the surface is not continuous with irregular nonwet areas exposed

Rosin bond

Bonding is achieved through a layer of solidified flux, usually rosin type. In its form, this joint has no metallic or electrical continuity, and has little physical strength

SG

Specific gravity

Tinning

Tinning of lead wires is a process of applying a thin coating of solder to the surfaces of the stripped lead. Wires are tinned to bond the strands together preventing fraying during handling and to preserve the solderability of the surface